DATE DUE			

SENTIENCE

SENTIENCE

by WALLACE I. MATSON

UNIVERSITY OF CALIFORNIA PRESS
BERKELEY   LOS ANGELES   LONDON

University of California
Berkeley and Los Angeles, California
University of California Press, Ltd.
London, England
Copyright © 1976, by
The Regents of the University of California
ISBN 0–520–09538–3
Library of Congress Catalog Card Number: 75–3774
Printed in the United States of America

Contents

Man need not be degraded to a
machine by being denied to be
a ghost in a machine. He might,
after all, be a sort of animal,
namely, a higher mammal.
There has yet to be ventured
the hazardous leap to the
hypothesis that perhaps
he is a man.

RYLE

Acknowledgments

To the National Endowment for the Humanities for unbureaucratically administered support of this project.

To Renford Bambrough, the Master and Fellows of St. John's College, and the many other friends in Cambridge who provided agreeable compensations for the agonies of composition.

To the Humanities Research Committee of the University of California for leisure to write the last chapter.

To Fred K. Beeson II and Charles Jarrett for acute, patient, thorough and constructive discussions of the work from conception to completion. Moreover, in Charles I found a research assistant on whose judicious assessment of the literature I could rely absolutely.

To Rick Doepke, William Gean, and Gonzalo Munévar for helpful comments on various drafts.

To Hubert Dreyfus, who first interrupted my analytic slumber.

To Bruce Vermazen, who gently restrained me from putting concepts into orbit.

To Warrington Colescott for Figure 1.

To Gene Thompson for the joke discussed in Chapter 5.

To Savannah Ross, Katy Dreith, Julie Martinson, and Yulia Motofuji, who were so nice about the typing.

To my wife, on general principles.

Finally, to Benedictus de Spinoza, who I believe had the truest conception of man.

I

PROBLEMS

"Man" said Democritus "is what we all know about."[1]
Strange words to come from the father of materialism, a
philosophy that might be self-evidently true if only
there were no people.

In a way, we do know about man: We are men, so we
know what it is to be men, as cats know what it is to be
cats. We know what we do, how we do it, and why, at
least sometimes. Wanting this or that, we size up our
situation, devise our plans, and act accordingly. This
pattern and its endless variants are indeed so familiar
that the natural human tendency is to deem anything at
all explained just insofar as it has been comprehended
in these terms. Hence the religious world view.

We have the advantage over cats of being able to
reflect articulately on our selves and our doings, to con-
ceptualize ourselves. It is an advantage, for it can lead to
knowledge even though theories are more likely to be
wrong than right. Theories about man tend to split him
into two parts, a grosser and a subtler. The most naive
of these dualisms is a simple application of anthropomorphic explanation to man himself. In keeping with
the general tendency to explain happenings by postulat-
ing a manlike agent, a man's behavior is theorized to be

1. Fragment 165 (Diels).

1

due to a little man inside manipulating the grosser limbs, tongue, etc. We smile at the *ka* of the Egyptians, but as we shall have occasion to note hereafter, his progeny are not extinct.

Under the influence of meditations on the difference between the foreman or king, who sits and gives orders, and the slave who need not think but only do his bidding, the model of man is refined into a ruling part, which does the planning and issues the orders, and the visible body, which is mere apparatus for carrying out directives—but alas, like slaves and subordinates generally, is apt to be sullen, lazy, rebellious, and the enemy of reason. Division and opposition are carried to the point where the ruling element is indeed thought of as completely separate from the body and alien to it, imprisoned in it as punishment for sin but to be freed on some glorious day.

In this kind of dualism, sentience is counted on the bodily side. It is too obvious that we see with our eyes and taste with our tongues for the sense functions to be dreamed of as divorced from the flesh. Ancient speculation grapples, in vain, with the question of how the eye sees, more successfully with how the ear hears; but it does not at all perceive as a problem how it can be that any material structure can have sensations. In terms of current controversy, all the ancients must be counted as identity theorists: they held, or at any rate took for granted, that sensations, including pains, were bodily processes throughout. Only the intellect—judgment, abstract reasoning, whatever it is that we are intelligent and virtuous with—was identified with that separable ruling element.[2]

2. For a fuller account, see my paper "Why Isn't the Mind-Body Problem Ancient?" in Paul K. Feyerabend and Grover Maxwell, ed., *Mind, Matter, and Method* (Minneapolis: University of Minnesota Press, 1966).

Nevertheless, presages of the problems of later times are to be found in the earliest materialism. When the personal model for explanation is rejected uncompromisingly, as it is by Democritus, and in its place is put the model of causal interaction of material particles, the question leaps to prominence: What is to be done about the actual, undeniable manifestations of sentience in ourselves? If there is nothing, really, but atoms hitting one another in the void, then that is all *we* are, really. How are we to understand these collisions as aches and pains, colors and tastes? Democritus seems to have been on to an explanation of the right *type* when he tried to account for tastes in terms of atomic shape; I say right *type* because he explained (for example) not sour taste but the (event of) vinegar's tasting sour; it was due, he said, to the spiky shape of the vinegar atom—that is, the disposition of that atom to prickle the tongue.[3] However, this still leaves us with the unanswered—indeed, for Democritus unasked—question why prickling the tongue should be *felt* at all.

Generally, and I think rightly, Descartes is credited with having invented the mind-body problem as we know it. He did this by beginning his philosophy not with a fund of common knowledge about the world and man but with *what we can be absolutely certain of,* which in his opinion is only the existence of the thinking self. This means that everything whose being is exhausted in my present awareness is to be put in one bin, distinguished from those items, in a second bin, that could fail to exist, at least conceivably, without altering my consciousness. The first bin, then, contains all my thoughts, *feelings, aches, pains, and sensations*—put now on the side of the mind, contrary to ancient practice, and contrasted with "the physical" which includes my

3. See Theophrastus, *De Sensu*, 65.

body, but only as a thing in itself insentient. Thus, there are two kinds of things: mental things and physical things. The one kind can exist and be just what it is without the other (as dreams are supposed to show), and indeed the major problem bequeathed to the new philosophy is how there can be conceived to be any intercourse between them at all.

It has come about, I know not how, that this division and the problems it engenders seem natural and uncontrived to people otherwise not much inclined to give credence to the extravagances of metaphysics. For example, half a century ago E. D. Adrian (later Lord Adrian and Nobel laureate), expressed a typical puzzlement at the beginning of his book on *The Basis of Sensation*:[4]

> The final chapter deals briefly and timidly with the relation between the message in the sensory nerve and the sensation aroused in our consciousness; briefly because the relation is simple enough in a way, and timidly because the whole problem of the connection between the brain and the mind is as puzzling to the physiologist as it is to the philosopher. Perhaps some drastic revision of our systems of knowledge will explain how a pattern of nervous impulses can cause a thought, or show that the two events are really the same thing looked at from a different point of view. If such a revision is made, I can only hope that I may be able to understand it.

My aim in this book is to persuade you that "the two events" are indeed "really the same thing"—to return us, so to speak, to archaic innocence in our view of man, but without jettisoning whatever sophistication we may have picked up along the way. That is, I want to present a *conceptualized* identity theory of mind and body.

I am not alone in having this goal, which was never

4. (London: Christophers, 1928).

lost sight of entirely: No less a philosopher than Aristotle pursued it in antiquity, while as for the heroic age of modern thought, if only people would read their Spinoza, there would be little need to say anything further. In our own day, dualist views have recently gone so out of fashion that I shall not pause to criticize them here.[5] And anyone writing after Ryle can hardly hope to do more than tidy up.

In carrying out this humble though, I hope, useful task I shall assume that a living man is a certain quantity of matter, a selection out of the ninety-two natural elements, combined in a complicated organization that functions as a unity in growth, nourishment, reproduction, and action. He lives for a time; sooner or later the unitary functioning ceases. When it does, that is the end of the man; what is left over is functionless matter which unless artificially preserved soon undergoes chemical decomposition and merges into the relatively undifferentiated reservoir of organic stuff—the dust. This view does not deny that a man has a mind or, if you prefer, a soul; it does deny, however, that "mind" or "soul" is a name for anything other than an aspect of the workings of the body.

This view is really a framework thesis, not a detailed theory, so most of the discussion can be carried on in terms of a simple, sloganlike statement:

Sensations are necessarily brain processes.

Or somewhat more exactly,

It is possible that sensations are brain processes; and if sensa-

---

5. Anyone requiring argumentation may turn to Lucretius, Book III, 450–850, whose twenty-odd arguments are still cumulatively conclusive.

tions are brain processes, then sensations are necessarily brain processes.

Except for modality, this thesis is the same as the contingent identity theory associated especially with Place, Smart, and Armstrong. Thus much of the work of showing its internal consistency is something that need not be done over again. The discussion can concentrate on the question of the modality in which I want to express the thesis: the claim, that is, that if sensations are brain processes then they cannot be, could not be, could not have been, anything else. This is the hard materialist line. It is traditional and also, it seems, hard to swallow, inasmuch as the resurgence of materialistic philosophy of mind seems to be due to its having been softened into the contingency view.

To many readers, especially perhaps those whose interests are more in psychology than philosophy, the question of the modality in which the identity theory is to be expressed will be of little concern. I hope they will not put the book down but will simply skip Chapter II. I have felt it necessary to include that chapter, however, for three reasons. First, I think the contingent identity theory is false, because it rests on a wrong theory of language. Second, if the contingent identity theory *is* true, there can be no reason to believe that it is true. For that theory is to the effect that while as a matter of fact sensations in *this* world are nothing but brain processes, there *could* be worlds in which sensations had no sort of connection with brain processes (or heart or liver processes for that matter). But if so, there can be really no way of telling that this world is not one of those worlds. It is not, nor could it be, a "scientific discovery" that sensations are identical with brain processes as a matter of fact, and as opposed to being ontologically different occurrences merely correlated with them.[6] The third reason will appear when in the final chapter it is shown that the defense of freedom against objections based on determinism depends crucially on there being a necessary identity.

After this part of the argument, the remainder of the book will be concerned with answering the broad questions "What is sentience? Why is there any such phenomenon? What difference does it make?" These vague questions will be sharpened up when the time comes to do so. My aim is to steer a materialist theory of mind along the strait and narrow path between two bogs, on the one hand epiphenomenalism, the view that everything would go on just as it does even if there were no consciousness in the world; on the other, (Watsonian) behaviorism, the paradoxical denial that there really is any such thing as consciousness.

In the third chapter I shall try to show that sentience is not to be conceptualized as a privileged view of a set of private objects. The conclusion is hardly new, but perhaps some of my reasons supplement and strengthen those urged by others. Chapter IV is intended to relieve the anxiety of those who fear that if the mind is identified with the brain, then man is nothing but a machine, to wit, a digital computer. Here I rely heavily on the thought of my colleague Hubert Dreyfus. Having said what man is not, I try in the succeeding chapter to

6. Indeed, there is something unconvincing about the insistence of the contingentists that "Sensations are brain processes" is a "scientific discovery." Who made it? When? How? There may be answers to these questions—though it is doubtful if anyone knows what they are—if the sentence is understood as filled out with "and not heart or liver processes, as we formerly believed." But that is neither here nor there in the philosophical controversy.

say what he is: to face the naive-sounding but I believe valid and important question, "What is the use of sentience? What can a sentient being do that a nonsentient being could not do?"

Lastly, I shall try to succeed where Epicurus failed and give a satisfactory materialist account of free will—one that does not merely reassure us that we are free as water is free to run downhill.

I think that what I am doing in this book is properly to be called philosophy, though I could face with equanimity an official ruling that it is not. But perhaps some apology ought to be made for dealing with questions that at least closely border on physiology and empirical psychology, subjects in which I claim no competence. Fortunately, the whole topic of sentience divides itself rather sharply into two parts. The neurophysiologist certainly does not wait to hear from the philosopher before proceeding on whatever line of investigation he is interested in. Nor should he. If he is, say, looking into the mode of action of the receptors for the sense of smell, all his concern is with what goes on when effluvia from a rose enter the nostrils and make contact with the specialized cells embedded therein. If he can give a satisfactory account of the energizing of the whole path from nostril to olfactory cortex, he has done his job. That is what smelling is, he is entitled to say. He will have discovered that when such-and-such events go on in the brain, such and such olfactory sensations are experienced. The question whether the olfactory sensation is the same as the neural excitation, or a ghostly accompaniment, is one in which the physiologist no doubt will have an interest, but not a professional interest. There is nothing he can do with his ingenious apparatus and techniques to resolve it.

The philosopher, on his side, who is interested in resolving this kind of question, need not know anything about the physiological details, as long as he knows that what occurred was the transmission of a message from a point on the nervous periphery to a region more centrally located. We know now that the transmission is electrochemical. Descartes thought it was hydraulic; Aristotle didn't think it happened in the brain at all, but in the heart. None of this bears on the philosopher's task, which is that of trying to make sense of the very notion of a relation between what initially impress us as being such different affairs that they could hardly have anything in common: the propagated physical change, on the one hand; the feeling, on the other. The philosopher's job will be finished when he can rightly claim to have given an intellectually satisfying account of the structure of a whole, which includes both the brain events and the feelings and which either leaves no gaps between them or, if it does, makes it clear why there must be gaps.

II

IDENTITY

I shall discuss the identity theory of mind and brain in the form that has become, as it were, canonical:

Sensations are brain processes.

This perhaps does not look like much in the way of a theory, and it isn't. Here at the beginning we cannot and need not make it more precise. Nevertheless, we should be mindful of the qualifications that have to be understood as applying to this slogan and the short-comings it retains. First the "are" in it signifies not coex-tension but class inclusion; no one wants to maintain that all brain processes are sensations. The meaning is that out of the indefinitely large number of happenings that occur in the brain, some, doubtless a small fraction, are sensations. Second, the word "sensation" is here just as inadequate as it is elsewhere in philosophy, where it is made to serve as a label for any arbitrary bit of consciousness. It suggests that seeing a tomato is somehow like feeling dizzy, or anyway to be put in some nearby philosophical pigeonhole. Perhaps even worse, its use encourages the groundless assumption that there is such a thing as a "mental individual" and that all these individuals—thoughts, feelings, "cona-tions," and whatnot—can be grouped as species of the one genus. However, being no more capable than Ryle was of finding a better term, I shall use it, exhorting readers and myself to wariness.

In the company of so shifty a linguistic character, it is a positive advantage to have as partner another vague phrase like "brain process." By this is intended any change, that is, redistribution of matter or energy, and especially those microprocesses of electrical conduction along the dendrites and through the synapses, as well as the electrochemical changes that facilitate or inhibit synapse crossings. But as already remarked, the exact or even approximate nature of these processes is not here in issue, and if they were passages of the animal spirits through little valves, it would be all the same. Further, we must give even "brain" a wide sense, to include at least the whole nervous system on occasion, not merely the part inside the skull.

The content of the slogan, such as it is, is programmatic, amounting to no more than the assertion that it is in principle possible to understand what are now referred to by such words as thoughts, feelings, and the like, in terms of an account the vocabulary of which consists exclusively of words signifying brain processes. But I have a hunch, shared by others who have concerned themselves with these matters,[1] that if such an understanding is ever achieved, its formulation will be a far more complicated affair than the provision of a dictionary with words like "feeling" or "thought" or even "pain of 3.89 microsades intensity" on the left, followed by descriptions of the brain processes in question. There is no reason to suppose that the words in which we ordinarily talk about our aches, feelings, and

1. For example, Thomas Nagel, "Armstrong on the Mind," *Philosophical Review*, 1970, especially p. 398ff.

insights have any intelligible one to one correspondence with definite and discriminable brain processes. Nor, as far as I can tell, is there reason to suppose that the inventory of basic principles of brain operation is now complete. For a successful reduction of sensation sentences to brain process sentences, we may need new concepts in both kinds. Common language terms like "belief" and "desire," though they can refer to brain states, would not be perspicuous when so used—just as "cake" is not a suitable concept for use in chemistry, even though cakes consist of nothing but substances listed on the Periodic Chart. If we want to describe in a rigorous chemical way the baking of a cake, we will not take the cookbook recipe and translate each term into its chemical equivalent; rather we will employ intermediate notions such as "carbohydrate."

Some further historical analogies will make my meaning clearer. There was the failure to reduce electromagnetism to physics before getting the concept of the field: When the only physical language available was that of tugs and shoves, these "strange kinky forces" were intractable. Likewise doomed to failure were attempts to reduce chemistry to physics before the development of models of the atom which exhibited it as complex, with separable parts, despite its name. Anyone who reasoned a priori that at bottom chemical phenomena could be nothing but mixture would have been right, but only in a trivial and unenlightening way.

However, pessimistic reflections of this sort can be overdone. Merely showing that it makes sense to speak of the earth moving around the sun would have marked an important conceptual advance, even in the absence of detailed astronomical data and proper vocabulary.

The identity theory is the contention that when the brain physiologist has done all he can, there is nothing more to do. The philosopher, then, is in Locke's image the underworker clearing away the rubble, by which polite metaphor is meant the perversions of linguistic usage perpetrated by philosophers of other persuasions. Undoubtedly there is much truth in this view, which may even be the whole truth. The master muddle that is referred to as the mind-body problem must consist in large part, if not wholly, of subsidiary phraseological tangles. To say so is not to take a doctrinaire stance to philosophy in general. It is but to make the platitudinous remark that in this philosophical area attention to linguistic propriety is desirable.

Now, in considering the debate between monists, those who hold that the mind is nothing but the functioning of the brain, and dualists, who say that mind is something over and above the merely physical, it is natural to try to apply the maxim formulated by Frank Ramsey and emphasized by John Wisdom that settlement of an inconclusive dispute is to be sought by denying the premise that both opposed parties accept. One recent writer, Daniel Dennett, has sought to escape being a dualist without having to allow the bleak paradoxes of the identity theory by following this procedure. He has diagnosed the shared premise as: "Mental terms, and also physical terms, refer to existent things," and he has denied it.[2] It will be instructive to look at his ingenious approach.

Dennett points to the existence in the English lan-

---

2. *Content and Consciousness* (London: Routledge & Kegan Paul; New York: The Humanities Press, 1969), a brilliant book, without which this one would have to be longer than it is.

guage of some nouns, which he calls "degenerate," that occur only in a few set phrases not allowing the inferences one can make from expressions of the same type containing normal nouns. For example, "sake" is found only in phrases of the form "for the sake of. . . ." From the fact that I did something for your sake it does not follow, or even make sense to say, that there exists a sake which is yours and for which I did something. This is extreme nominal degeneracy. Intermediate between sakes and tables Dennett discerns a number of nouns such as "mile," "degree Fahrenheit," and "voice," which occur in a variety of contexts but nevertheless are, like "sake," nonreferential—not in the sense that "centaur" is nonreferential, namely, that there do not happen to be any centaurs; but that the idioms do not license inference to expressions of the form "There is a . . . which. . . ." For example, from "There is a mile between our houses," we cannot proceed to assert "There exists a mile and that mile is between our houses."

Dennett claims that "voice" is a noun of this kind. We can lose and regain our voices, a voice may survive the speaker's death on a phonograph record, there may be a voice in the dark; for all that,

A voice is not an organ, disposition, process, event, capacity, or—as one dictionary has it—a 'sound uttered by the mouth.' The word 'voice' as it is discovered in its own peculiar environment of contexts, does not fit neatly the physical, nonphysical dichotomy that so upsets the identity theorist, but it is not for that reason a vague or ambiguous or otherwise unsatisfactory word. This state of affairs should not lead anyone to become a Cartesian dualist with respect to voices; let us try not to invent a voice-throat problem to go along with the mind-body problem. Nor should anyone set himself the task of being an identity theorist with respect to voices. No plausible materialism or physicalism would demand it. *It will be enough if all the things we say about voices can be paraphrased into, explained by, or otherwise related to statements about only physical beings.*[3]

Dennett believes that "mind" and the words of the associated vocabulary are like "voice," so that we can similarly avoid both dualism and the identity theory.

This is puzzling. To say of minds that "it will be enough if all the things we can say about [them] can be paraphrased into, explained by, or otherwise related to statements about only physical things" is, to my understanding at least, precisely and concisely to state the aim of the identity theory.[4] If you can get away with paraphrasing the statement that you have a pain in your left big toe into something like "there is occurring a stimulation of the C-fibers," the field is won; for this paraphrase is simply a particular application of the general rule that pain statements are translatable without remainder into brain statements, which is the cash value of our slogan "Sensations are brain processes." But the identity theorist has an advantage over Dennett; he can say, with Ryle, that minds and pains, as well as voices, exist, whereas Dennett is forced to claim paradoxically that the right answer to "Do voices exist?" is "No." But it is hard to understand what is supposed to be gained by this aberration. If it is discounted, as it should be, Dennett turns out to be an identity theorist *malgré lui*.

We should share Dennett's thankfulness that there is no voice-throat problem. However, it may be interesting to consider what it would be like if there were one.

---

3. *Op. cit.*, p. 9; emphasis supplied. Quoted by permission of the publishers.
4. Unless "or otherwise related to" is the escape hatch. However, Dennett does not seem to use it as such.

We can, I fear, readily imagine philosophers saying things like these:

1. This statement, "I talk, I speak," must infallibly be true every time I utter it. But I can imagine that I have no throat.

2. The throat is located in space and time; the voice, only in time. (If someone objects, "But the voice comes from right here"—tapping his throat—he will be asked, in a superior tone, to calculate the number of cubic inches occupied by Pericles's Funeral Oration.)

3. "Voice" certainly doesn't mean "vibration of vocal cords" since one can correctly use the word "voice" without knowing anything about anatomy. Therefore, voices are quite distinct from vibrating vocal cords, therefore separable from them.

4. Besides, voices can only be heard, whilst vibrating vocal cords can only be seen or felt.

5. And someone looking through the autolaryngoscope would simply see muscles shaking—he would not see his voice; indeed it is nonsense to talk of seeing a voice.

6. —As is shown also by the fact that one can ask questions about voices that are nonsense when asked of vibrating vocal cords, and vice versa: of the former, are they loud or soft? high pitched or low? of the latter, are they fast or slow? tense or slack?

7. Vibrating vocal cords are merely matter in motion whilst voices are intentional—they bear meanings.

8. And while, in a loose sense, both you and I can hear my voice, in philosophical strictness, only I can; the real voice that I hear is altogether different from the spurious one "out there" which exists on phonograph records and is mere appearance. We are isolated voices, each one within his own ears.

9. Therefore, voices and vibrating vocal cords are distinguishable, separable, and related, if at all, by mere correlation.

Here the philosophers split into sects: parallelists, interactionists, epiphenomenalists, occasionalists, double-aspectists, positivists, *et al.*

After this had been going on for some centuries, some heretic might dare to assert that notwithstanding these arguments, the voice just is nothing but vibrating vocal cords. He would take care, however, to soften the paradox by adding the qualification that the identity is merely contingent. For (he might add) it's a great scientific discovery that the voice is the vibrating cords; and all scientific discoveries are of matters of fact, which are always described by contingent statements. It could have been otherwise. A disencorded voice is logically possible.

If we are not so far gone in philosophical reverie as to fall for the voice-throat problem, why not? I think we would want to say something like this: The word "voice" got into the language before the phrase "vibrating vocal cords" because the former is a label for something we encounter all the time, whether we are educated or not; the meaning of the latter has to be learned and is based on discoveries in anatomy. The former term is retained because of its usefulness for talking about aspects of vocal behavior that we often do want to talk about; it fits neatly into our speech patterns and interests whereas a translation into the idiom of vibrating vocal cords would be long, clumsy, and pointless. Nevertheless, nothing can be said in the voice idiom that cannot also be expressed exactly in the terminology of vibrating vocal cords. For example, "He has a high-pitched voice" = "The average frequency of vibration of his vocal cords is somewhat greater than the mean average frequency for the adult male population." This does not specify the cause of his having a high voice; it states what having a high voice is. In a certain sense it may be said to explain the high voice—but that sense is only this: the vibrating vocal cord idiom is technical, that is to say, is embedded in a comprehensive acoustico-anatomical theory where the individual concepts are interrelated through entailments. "High-pitched voice" is a logically isolated notion taken from everyday vulgar speech; "high frequency vibration" virtually announces to us that it is extracted from a theory.

Philosophers may be tempted to suppose that at any rate there is this much to the voice-throat problem: that "voice = vibrating vocal cords" is only a contingent identity. They will point out, rightly, that we know that vocal cords aren't necessary for voice production. We can produce a voice by various means, as is shown by devices actually used to "give voices" to persons who have had their vocal cords surgically excised.

This does prove that voices do not have to be vibrating vocal cords, but that fact is irrelevant to the main point, which is not concerned with the accidental features of the materials in which a certain structure is realized, but with the structure itself. Voices do not have to be vibrating vocal cords, but they do have to be vibrating *somethings*. From the possibility of voices where there are no vocal cords, we cannot deduce that there can be voices where there is no vibrating physical structure at all. We can build devices with structural similarities to the vocal cords. When we do, we have, in the relevant sense, a voice; and only then.

The moral for mind-brain identity is that in claiming

more than mere contingent as-a-matter-of-fact identity of minds and brains, we are not to be put off by the observation that it might turn out to be possible to build a thinking machine out of wires and plastics. Such a device would constitute not a refutation but a confirmation of the identity theory. "Brain processes" in the slogan must be understood as vaguely signifying processes in a certain kind of structure as yet only imperfectly understood. The materials out of which the structure is composed may be of various sorts, as chess is the same game no matter what the pieces are made of.

The voice-throat parody of the mind-body problem may help to make an important negative point. The astute reader will have noticed that the parallel lacked plausibility at two points, in attempting to provide analogues to the alleged privacy and intentionality of the mind (nos. 7 and 8). These are the features of "the mental" that most modern defenders of dualistic views rightly find crucial to the defense of their views. Because "the mind" does not here run in linguistic tandem with "the voice," we are not likely to be able to catch them in the same trap. We must also acknowledge that although in a way it hardly matters whether we say that voices exist or voices don't exist, it is otherwise with mind. There is something that begins in the morning and stops (or, at most, proceeds dead slow) when we go to sleep at night, and ends forever (or else doesn't) when we die. No doubt there are grave defects in the descriptions of this something; they may be so serious that it should not even be called a something; but as Wittgenstein observed, it's not a nothing either. It exists.

If Dennett has not found a middle way between dualism and identity theory—and I do not know of anyone else who has even tried[5]—the next step is to inquire into the consistency of the identity theory. We must begin by deciding what the logical form of the theory is or ought to be. Is "Sensations are brain processes" analytic or synthetic? a priori or a posteriori? necessary or contingent?

There is a preliminary question: To answer these three questions, do we have to make three investigations? Or two, or only one? Many would say: Just one. For it is common to treat "analytic," "a priori," and "necessary" as synonyms. This practice is supposedly justified by uncontroversial arguments due mainly to David Hume. But let's see.

A priori/a posteriori, though originally referring to reasoning from causes to effects and vice versa, has since the eighteenth century been taken as an epistemological distinction. What is knowable independently of sense experience is a priori, prior, that is, to experience; whereas what can be known only by sense observation or other experience is a posteriori. Thus, all of mathematics is a priori knowledge since we never have to perform any experiments in order to prove a theorem; on the other hand, it is held that all of physics, save the ancillary mathematics, is a posteriori, for there is no way of knowing what the world is actually like without looking at it and testing it.

Analytic/synthetic is a distinction of logical forms. A proposition is said to be analytic if the predicate analyzes the subject, that is, if the concept which is the predicate turns out upon analysis to be the same, or part of the same, concept as that which is the subject of the sentence. The sense of "analysis" used here depends

5. I count Ryle as an identity theorist, though he would no doubt protest.

upon the notion of words as signifying concepts which are either simple or capable of being taken apart into at least relatively simple components. A definition is such an analysis. Thus, "bachelor" is supposed to signify a complex concept having the two components "male" and "unmarried." Hence, the proposition "All bachelors are unmarried" is analytic because the predicate term "unmarried" is part of the analysis of the subject term "bachelors." A proposition in which this analysis cannot be carried out is synthetic. "All bachelors are wealthy" is synthetic, for whatever the analysis of the concept of wealth may be, it does not coincide either wholly or partly with that of bachelorhood. Being wealthy is no part of the very notion of being male, nor of being unmarried.

The distinction as thus defined applies only to propositions in the subject-predicate form. Originally, this was not thought to be a restriction on its scope since all propositions were supposed to be translatable into the subject-predicate form. Since that doctrine has been abandoned by logicians, the definition of "analytic" has been relaxed, or enlarged, to mean "true by virtue of the concepts involved alone." Thus, "John is taller than Jane," said by Russell to be a proposition having two subjects and no predicate, is synthetic, it being no part of the concepts of John and Jane that they stand in any particular relation of height; but "A mile is longer than a kilometer" is analytic in this extended sense, for what we mean by a mile is a unit of distance that is greater than a kilometer. Even if it is not laid down explicitly in the definition of either unit that it is greater or less than the other, this consequence is deducible—derivable a priori—from their respective definitions.

This distinction is supposed to apply exhaustively

and exclusively: Every proposition is either analytic or synthetic, and no proposition is both. The reason is that every word, if it is meaningful, signifies a concept;[6] and that concept is either simple or analyzable into simpler subconcepts. A word such as "mistress" may change its meaning, that is, acquire or lose meaning components in the course of time; different people may use the same vocable to signify different concepts, as supposedly people west and east of the iron curtain have differing concepts in mind when they use the word "democracy." And some concepts—democracy, baldness, baroque—are vague, which is to say that it is difficult or even in practice impossible to be sure just what the necessary and sufficient conditions are for their application. But in principle, it is held, at least on any particular occasion of use, the concepts have their analyses, whether we can discover them or not.

The last of this triad of distinctions, necessary/contingent, is in its origin a metaphysical dichotomy. Aristotle defines "necessary" as "what cannot be otherwise." In his writings he applies the notion both to propositions and to states of affairs. He calls the proposition "Motion is eternal" a necessary truth, holding that it is not a mere matter of fact that motion continues forever, but something that in the nature of things could not be otherwise. This state of affairs, consisting in motion going on forever, is also called necessary. It could not be otherwise: Its denial expresses an absolute impossibility.

The opposite of "necessary" is "contingent." A state of affairs is contingent if it is not necessary, if it could have been otherwise, that is, its being or not being

6. With some exceptions (for syncategorematic terms, for example) that need not concern us here.

depends on conditions external to itself. It is a contingent state of affairs that there is a tree in the quad. It's there, all right, but if it hadn't been planted, and watered, and protected from lumberjacks, and so on, it wouldn't be there. Derivatively, for Aristotle, "There is a tree in the quad" may be said to express a contingent proposition.

Most modern logicians consider Aristotle mistaken in applying this distinction both to states of affairs and to propositions.[7] They hold that the distinction is modal and should be applied only to propositions. A proposition is necessary, or necessarily true, if and only if it is a member of all possible worlds. Otherwise, it is contingent. "Possible world" means any set of propositions that are logically mutually consistent, that is, no member of which is the contradiction of another member. The set may contain any number of propositions, or none. A proposition is said to be true in a certain possible world if it is a member of that set and the set is consistent. It is true in all possible worlds, that is, necessarily true, if every consistent set of propositions contains this proposition.[8] "If anything is square, it has four corners" is such a proposition. It does not assert that anything is square, nor indeed that anything has any property at all; it says only that if something has one property, it must have another; and these properties are such that anything having the one must have the other,

7. But it is strange to accuse the inventor of a distinction of having got it wrong.

8. All necessary truths are consistent with each other and with all other non-self-contradictory propositions. Thus they are members of every possible world. Since the set consisting of all and only necessary propositions contains no existential assertion, this set is said, somewhat curiously, to "describe the null world."

no matter what else may be the case—which is the same as saying that it is true in every possible world.[9]

These distinctions are very old, or at any rate the first and third are. Their conflation has proceeded this way:

The conclusion that all a priori truths are necessary was reached by the Greeks almost as soon as they had discovered the conception of mathematical proof. The poem of Parmenides (early fifth century B.C.) is mainly concerned to set off what can be known by pure reason, "well-rounded truth" that never changes, cannot be otherwise, from "guesswork" that derives from "the unseeing eye and the echoing ear."[10] Parmenides evidently also believed the converse, that all necessary truths are a priori; thus he held the necessary and the a priori to be equivalent. Plato followed him in this, and Aristotle followed Plato, though emphasizing more than Plato did the need for sense experience to provide the data which, when sized up properly, trigger the insight into necessity. Aristotle further asserted what Plato and Parmenides had denied, that there are necessary truths about the objects of sense perception.

The fact that the human mind is capable of grasping by its own unaided power not just the way things are but the way they have to be, struck the Greek thinkers and their Christian successors as a wonderful thing. The analytic/synthetic distinction was created to debunk these pretensions, or to put it more politely, to explain how a priori knowledge of necessity is possible. The

---

9. But evidently "Motion is eternal" cannot be necessarily true according to this definition. For there does not seem to be anything inconsistent about the set of propositions containing the one member "Nothing moves," and the union of this set and "Motion is eternal" is inconsistent.

10. Diels and Kranz, *Die Fragmente der Vorsokratiker*, 10th ed., 28B 1, 7.

story goes this way: All necessary or a priori truths, that is, all propositions which can be known independently of experience of the world, turn out on scrutiny to be analytic. The propositions of mathematics, for example, are deduced from axioms, taken to be self-evident necessary truths. But the process of deduction, according to the analytical view, is merely a concatenation of substitutions of terms that have the same meaning; and the axioms are themselves definitions—explicit statements of how the terms involved are to be taken. Thus, "If equals are added to equals the sums are equal" is not to be regarded as an insight into the nature of things but a remark about how the mathematician proposes to use the terms "equal," "add," and "sum." No wonder, then, that we can be sure that $2 + 2 = 4$; to simplify greatly, but not to misstate the essential point, it is because "2" means "$1 + 1$," and "4" means "$1 + 1 + 1 + 1$"; so the equation reduces to $1 + 1 + 1 + 1 = 1 + 1 + 1 + 1$. And identities such as $e^{\pi i} + 1 = 0$, while less obvious and more interesting, are no different in principle.

Necessity thus resides not in nature but in language. The things in the world, whatever they are, are "entirely loose and separate"; necessity holds only between concepts. Wherever we have two independent concepts or descriptions, and things that are described by them, we have two separate and independent things or at least we could have. It is a necessary truth that my sofa is identical with my chesterfield, but that is only because as I use the words, "sofa" and "chesterfield" are sounds or marks for the very same concept. On the other hand, if my desk is an old packing crate, the two separate concepts "desk" and "packing crate" merely happen de facto to refer to one and the same object. Because they are different concepts, the sentence in which they both figure is synthetic. Something could be an old packing crate without being my (or a) desk. For I could buy a Chippendale antique and discard this— which would then cease to be my desk, without ceasing to be an old packing crate, unless I smashed it or burned it.

This then is an example of contingent identity: logically independent descriptions applying to the same object. Now, how are we to tell whether two concepts are logically independent? The formal criterion is that it should be at least logically possible for something to be an instantiation of the one without being at the same time an instantiation of the other; or more formally still: A and B are independent concepts if and only if there is at least one possible world containing an instantiation of A but not of B, or of B but not of A. If we lived in a world in which all desks were in fact old packing crates, we should still recognize the identity as merely contingent because we should appreciate that there *could* be such a thing as a desk that was not an old packing crate. In our world, all rational creatures are featherless bipeds, but we are aware that there could be a rational being covered all over with feathers, like Papageno. So imaginability turns out to be the criterion of independence: two concepts are independent if and only if it is possible to imagine the one applying where the other does not.

This, however, is like the infallible direction for catching a bird by sprinkling salt on its tail. How are we to tell whether what we can in fact imagine is the same as what in an absolute sense it is possible to imagine? Some people are more imaginative than others. Probably there are people who, if asked whether they can imagine a rational being with feathers, would sincerely reply that they could not. So in practice it looks as if we should have to apply the criterion only one way, as a sufficient but not necessary condition; that is, if someone can imagine A without B, then A is independent of B. Unfortunately, even this may be too much; we seem to be able to imagine logical impossibilities such as time travel and circle squaring. We are fallible enough when it comes to knowledge; there is no reason to suppose we are better off in the faculty of imagination.

Another suggested criterion has it that two expressions must have different meanings, thus signify different concepts, if it is possible for someone to know the meaning of the one without knowing the meaning of the other—"knowing the meaning" generally being taken as amounting to knowing how to use the expression correctly. "Voice" can't mean "vibrating vocal cords," nor "water" mean the same as "H_2O" because someone might well know the meaning of the first term in each pair without knowing anything about the other.

Stated thus starkly, the criterion leads immediately to such paradoxical consequences as that no words in different languages can signify the same concept, nor indeed that there can even be such a thing as a pair of synonyms. For it is always possible for there to be someone who speaks the one language and not the other, or who is familiar with only one of a pair of synonyms. Nor could most mathematical equations express conceptual identity.

Perhaps it is possible to fix up the criterion to take care of these objections.[11] I shall not pause to investigate the matter, as I shall make more fundamental criticisms of the ideas behind it in what follows.

11. Though the prospect is gloomy. Evidently any program to have this result would have to begin by making a mysterious distinction between knowing the concept and (merely) knowing the expression, and by asserting the paradox that to know a concept is to know all its entailments. But if these moves are made, the criterion will already have degenerated to the triviality that if the concepts are different, then they are different.

If we accept received opinion on the distinctions as sketched above, we seem to be forced to the following conclusions about the mind-brain identity slogan. First, "Sensations are brain processes" can hardly be knowable a priori. Surely, if sensations are brain processes, the only way we can know it is by noting such facts as that particular sensations cease when particular parts of the brain are put out of commission and that sensations can be made to occur by artificially stimulating the brain, as by an electrode. By the anciently accepted equivalence of the a priori to the necessary, it follows that the slogan does not express a necessary truth. Further, it seems obvious that the concept of sensation, whatever it may be, is not the same as, nor included in, the concept of brain process. If proof is needed, it is easily supplied by noting the easy imaginability of sensations occurring in the absence of brains, or indeed of any bodies at all: We have no trouble in imagining our existence in a disembodied state. Therefore, by the argument connecting analyticity to necessity and the a priori, we once more reach the conclusion that the proposition expresses a contingent truth at most. Hence, there seems to be no doubt that the identity of mind and brain, if it is a fact, is a contingent fact. The identity theory can then be expressed as the contingent identity theory, and expressed as:

Sensations are brain processes, but sensations are only contingently brain processes.

This is the conclusion reached and defended by a group of philosophers including U. T. Place, J. J. C. Smart, and D. M. Armstrong.[12] But before we decide to adopt it, we need to pay attention to one or two matters.

12. U. T. Place, "Is Consciousness a Brain Process?" *British Journal of Psychology*, 1956; J. J. C. Smart, "Sensations and Brain Processes," *Philosophical Review*, 1959; D. M. Armstrong, *A Materialist Theory of the Mind* (London: Routledge & Kegan Paul, 1968).

One important respect in which "Sensations are brain processes" differs from "His desk is an old packing crate" is that the former is an expression of what the Australians call "strict" identity. The packing case takes on the role of desk, so to speak, and can drop it. But sensations, according to the theory, can neither become nor cease to be brain processes—brain processes are what they *are*. Even if in some other possible world there are sensations that aren't brain processes, still in this world, we are saying, sensations are the very same things as brain processes—in the way (to use an example of Smart's) that the number seven is identical with the only prime number between five and eleven.[13] But if so, then it might seem as if the sensations of this world could not possibly be anything but brain processes; for how could anything differ from what it is the very same thing as? What a thing is the very same thing as, is, simply, itself. But it could never be a merely contingent fact that a thing is the very same thing as itself. Yet we are assuming, for the sake of argument, that in some possible world, sensations are not brain processes. Now if this assumption is to mean anything germane to the point, it cannot signify only that some things very like terrestrial sensations are, somewhere else, not the same things as some other things very like terrestrial brain processes; on the contrary, it must mean that somewhere else, or even right here, things that *are* terrestrial sensations are not the same as things that *are* terrestrial brain processes. So we are saying that in this, the actual, world, sensations are the very same things as brain processes, but that in another possible world, these very same entities would not be brain processes, that is,

13. "Seven is identical to the only prime number between 5 and 11" is, of course, a priori, analytic and necessary. The claim being made is that there can be identity this strict that nevertheless is expressible in a contingent proposition.

would not be what they are the very same things as. This looks like a contradiction. I shall try to show that it is—and, consequently, the identity of mind and brain, though a scientific hypothesis, is a necessary truth if true at all.[14]

Proponents of the contingent identity theory content themselves with offering examples to prove their thesis that identity can be at the same time strict and contingent; a legitimate procedure, if the examples offered really do exhibit both properties. The list of examples offered is fairly short and includes:

1. The morning star is identical with the evening star.
2. Lightning is an electrical discharge.
3. The gene is the DNA molecule.

Let us examine these.

"The morning star is the evening star" is supposed to be an assertion of strict identity, for each is nothing more and nothing less than the planet Venus. The phrase "the morning star" denotes the planet Venus, and so does "the evening star." Yet the identity is contingent; things could have been otherwise inasmuch as we can easily imagine another constitution of the solar system in which the two phrases would denote

14. See my paper "Against Induction and Empiricism," *Proceedings of the Aristotelian Society*, 1962, for an earlier claim that the natural sciences traffic in necessary truths.

The remainder of this chapter is a reworked version of "How Things are What they Are," *The Monist*, 1972. A very similar argument is advanced in "Identity and Necessity," Saul Kripke's contribution to the collection *Identity and Individuation*, Milton K. Munitz, ed., (New York University Press, 1971), and in "Naming and Necessity," in Donald Davidson and Gilbert Harman, eds., *Semantics of Natural Language* (Dordrecht: Reidel, 1972). Alvin Plantinga also has argued for natural necessity in "World and Essence," *Philosophical Review*, 1970. He purports to prove, with the aid of some implausible premises, that *all* properties that anything has, it has necessarily. One would be more comfortable in Place's frying pan than in Plantinga's fire.

distinct objects. So, while there is identity of reference, there is not identity of meaning. The fact that there is identity of reference was discovered empirically, as it had to be; no amount of meditation on concepts could have excogitated this truth.

So runs the doctrine, whose simplicity, clarity, and decisiveness are delusory. In the first place, the statement, as put, is not even true, for sometimes Jupiter, for example, is the evening star. But that is a minor point. There is a true identity signified by "Hesperus is the same as Phosphorus." The ancients noted the presence, sometimes, in the evening sky of a bright planet, which they named Hesperus. They noted also the presence in the sky before dawn, sometimes, of a bright object which they called Phosphorus. Later, the Pythagorean astronomers announced from their observatory that these were not two distinct planets but one and the same, the one that we call Venus.

A scientific discovery, not a conceptual analysis. But how was it made? First, Hesperus itself had to be identified, not simply as "the evening star"—it was not always in the evening sky, and when there, not always the brightest—but as a planet with a certain path plotted against the background of the fixed stars. Similarly for Phosphorus. The realization that in truth there was only one planet, not two, came when it was noticed that the extrapolation of the one path coincided with that of the other and that, moreover, the interval between the disappearance of Hesperus from the evening skies and the appearance of Phosphorus in the mornings was just what was to be expected for the regular progress of a single planet on a single orbit—in other words, that the observations could be accounted for by the assumption of a single body; moreover, the assumption of two bodies was shown to be untenable, for those bodies would have had to be located in the same place.

About anything whatsoever that is perceived on two or more separate occasions, we can raise the question whether it is the same thing or two similar things. Is the morning teapot or cat or wife the same as the one we see in the evening? Usually yes; conceivably no. And if we have to make very sure, we will scrutinize the object carefully, call on other witnesses to testify about what happened during the time we were away, perhaps look at photographs made in the interval, consider the likelihood of anyone's having a motive for making a substitution, and so on. This shows that our *knowledge* that the object in the evening is the same as the object in the morning is a posteriori. However, we cannot conclude merely from this that the proposition "This evening teapot is identical with the morning teapot" is contingent, unless we accept the equation between what we can have a priori knowledge of and what is necessarily so. And on what does that equation rest? It is true for mathematics, no doubt; but its extension beyond mathematics is problematic, resting perhaps on no more than a certain feeling, which some Greeks had, that it would be undignified to have to bustle around like Herodotus to garner truths about the ways things *must* go.

Whatever can be identified can be misidentified. So there is always the logical possibility of error when one attaches any label to anything in the world: "Hesperus," "Dinah," "wife." We could, in each case, be deceived, in the weak sense that it could never amount to a contradiction to say that we had mistaken a substitute for the real thing. But even what would ordinarily be regarded as assertions of self-identity are open to risk in this respect, hence a posteriori: "Hesperus = Hesperus"[15] is just as risky as "Hesperus = Phosphorus"; "This is the same ache I woke up with" is as shaky as "This ache is a stimulation of the C-fibers." But if in fact we have not made a mistake, it is still an open question whether what we said was necessarily or only contingently true.

We are tempted to think that even though Hesperus is the same as Phosphorus, they might have been different. We imagine another sort of solar system, in which there is one planet, which we name Hesperus, shining brilliantly every evening for some months; and another planet, named Phosphorus, shining in the mornings, really quite distinct from Hesperus. Well, how distinct? On an orbit rather like the one we now calculate for Venus? Or exactly like it? If the former, our supposition is reduced to the triviality that things could be different and still look rather like the way they do now. If the latter, we must come at last to the assertion of the bare logical possibility, if it is even that, of a body's being miraculously annihilated, with simultaneous creation of another body, just like it only numerically different, in the very same place.

To return to what actually happens: The astronomers develop a theory, which they call a theory of planetary motions. It is based on positional observations of the planets and has two purposes: to make possible the prediction of further positional observations and to

15. Without explicit or implicit reference to an identical time and occasion of observation. "Hesperus observed by me at 0600 GMT = Hesperus observed by me at 0600 GMT" achieves exemption from possible error, if it does, only at the price of saying nothing. To point to something and say "This is identical with this" is strangely thought by some philosophers to be a scheme for making an absolutely foolproof assertion. On the contrary, it is not an assertion.

explain the observed positions. It does the explaining by showing how by making certain assumptions about relative motions of planets and earth, the paths of the planets can be conceived as simpler than direct observation suggests when naively interpreted.

However, we must not be misled about the nature of the theory itself by the circumstance that it is developed in close conjunction with a mass of observational data and with a view to being applied to those and to further data of the same sort. The theory itself is just a body of abstract hypothetical propositions tied together by logical entailment. A set K of elements p_1, p_2, . . . p_n is defined by assigning to these elements certain definite properties, such as that of attracting each other inversely as the square of the distance between them. From these definitions, other propositions are deduced. The theory is pure mathematics, consisting solely of definitions and their necessary consequences. The propositions composing it are a priori, analytic, and necessary. It is entirely general and does not per se apply any more to the solar system than to an infinitude of other "possible worlds."

But it can be—and was designed to be—applied to the solar system. This application is accomplished by identifying the elements of the set K with the actual planets, introducing certain numerical constants into the equations, constants that are not themselves part of the abstract theory. In this way we leave the realm of purely conceptual relations and take up the risky business of making assertions about the way the world is. We say, "Venus has such and such an orbit." But a more perspicuous expression, relating the actual planet to the theory, would be put in hypothetical form: "If that object which we call the planet Venus has the properties that define the element p_2 of the system, then. . . ." This whole hypothetical proposition is also a priori, analytic, and necessary. But when we commit ourselves and say, categorically, "The planet Venus possesses the properties defining p_2," when, that is to say, we start to say something about an object in the world that just might spite us by not having the properties we think it has—then, but not till then, we have climbed out on to the a posteriori limb. The only way to find out whether Venus has the properties of p_2 is to look.[16]

How does the assertion "Venus has the defining properties of p_2" fit into the threefold classification of propositions? Well, it is surely a posteriori, as we have just pointed out. But is it analytic or synthetic? We may be tempted to say that it is certainly not true by definition; therefore it must be synthetic. But this would be over hasty. "p_2" no doubt signifies a concept—there are definite necessary and sufficient conditions for being p_2—but what about plain pre-astronomical "Venus"? This term signifies only something like "that bright thing up there," an expression which hardly refers to a concept in a sense of "concept" that allows us to talk of necessary-and-sufficient-conditions-for-being-Venus. It is a label or particular demonstrative. We must not assume, then, that the proposition under consideration can be classified as either analytic or synthetic. For analyticity is a relation that can only hold or fail to hold between two concepts; and we must not assume that at this stage we already have two concepts.

---

16. The fact that actually the looking was done long ago—that the theory was worked out after we had looked, and with the help of the data got from looking—obscures the logical relation between the theory and its application and gives rise to much confused talk about "induction."

Necessary or contingent? If we suspend judgment on whether being a posteriori negates being necessary and on the question whether the proposition is analytic, then this question is premature. Furthermore, to decide whether "Venus = p_2" is necessary or not amounts to deciding whether, granted that Venus is p_2, it "could have been otherwise." And this phrase has not yet been adequately scrutinized.

Let us call the planet Venus, that bright thing up in the sky, "Venus-W" (for "the Venus of the world"). Let us call by the name "Venus-θ" (for "the Venus of the theory") what we have hitherto been referring to as p_2, that element in the set K which we want to associate with, or interpret as, Venus-W. Now the propositions within the theory, which involves only Venus-θ, are all of them necessary, as we have seen. And they assert nothing about the world. Likewise for conditional propositions which begin "If Venus-W = Venus-θ . . ." If the theory turns out to be "wrong," these conditionals are vacuously satisfied, and their necessary truth is not affected, though in that case they, and the whole theory, would not be very useful. The theory's being "wrong" or "false" can consist only in the identification "Venus-W = Venus-θ" being incorrect.[17]

The sense given here to "identification" is as follows. Venus-W is a label which we affix, we hope consistently (that is, we hope we name one and the same object by it every time we use it), to a massy object revolving around the sun. The object thus labeled has indefinitely many properties, a few of which we know, such as a dense atmosphere and a high surface temperature; and

17. Aside, of course, from internal inconsistency.

many more that we do not know. Venus-θ, on the other hand, is a constructed notion, a concept. There are necessary and sufficient conditions for being correctly referred to as Venus-θ. These are explicitly set out in the theory. They consist mostly of positional specifications. The identification of Venus-W with Venus-θ, then, is the assertion that the whole set of properties that define Venus-θ is realized as a subset of the properties of Venus-W[18]—which, however, before its identification with Venus-θ did not have any *defining* properties.

Let us compare the logical situations before and after the W–θ identification has been made. When we are dealing only with Venus-W, we get a fix on it and write in the notebook: "At t_k, Venus so many hrs min sec rt ascension, so many deg min sec altitude." This report of observation is a posteriori, synthetic, or, at any rate, nonanalytic if only because, as we have seen, there is no concept to be analytic with; and—if some tyrant forces us to decide between necessary and contingent—contingent. For at this stage, if asked what we mean by Venus, we can only reply: "That there," and point at it. We mean the prominent feature in the sky in the direction we're pointing; and we wouldn't be deterred from giving this reply even if the bright feature were a bit over to one side or the other from where we expected it to be. Certainly it is contingent by the imagination test, if we think that worth applying. There are possible skies of possible worlds that have their bright features differently distributed.

---

18. Together with the implicit assumption that none of the additional properties of Venus-W is capable of neutralizing, so to speak, any of the Venus-θ properties. For example, there mustn't be a volcano on Venus-W capable of acting as a jet engine and pushing the planet out of its orbit.

After we have the theory, hence are dealing with Venus-θ, the situation is very different. In the first place, the exercise we have been imagining would be pointless, or, rather, if we did take a fix on Venus, it would not be in order to find out where Venus is—the theory and the almanac calculated from it tell us that— but either to calibrate our instruments, or to find out where *we* are, or perhaps to test the theory. For now it is one of the defining properties, or an entailment of them, that Venus will be—*must* be—at such a place at such a time.

When we have started talking this way, we have witnessed the birth of a third and last Venus, Venus-ϕ (for "Venus of physical science")—in garbled mythology, a mortal fusion of the heavenly Venus-θ and the Venus-W of the people. Put in less poetical words, we now have a term that combines the mere designatory or pointing out function of the W term[19] with the conceptual content of the θ term. People who use the ϕ term at once signify by it the concept expressed in the definition or scientific description and designate by it the object in the sky, thus committing themselves to the assertion that the theory really does apply. "Venus-W" means "that thing we've all agreed to call Venus"; "Venus-θ" means "the element p_2 in the mathematical theory of orbiting objects, when such and such values are assigned to the constants in the equations"; and "Venus-ϕ" means "The actually existing object that we call Venus, which has the orbit of p_2 with such and such values of the constants."

19. It is a baseless prejudice to suppose that only grammar book type "demonstratives" can have the function of pointing out without describing. Many common as well as proper nouns do. See below, p. 45 ff.

We can now sort out the types of sentences into which these three kinds of terms enter.

1. All nontautologous propositions ascribing properties to the designata of W terms, or relating W terms to other terms,[20] are a posteriori and synthetic. They are subjectively contingent, that is to say, the user of the W term cannot give any reason why the proposition he utters must be true, why the thing he is talking about could not be otherwise. Whether the proposition is objectively contingent or necessary, is a question that cannot be answered without giving a theoretical account of what the W term signifies—but then we no longer have a W term, we have a ϕ term.

These generalizations have some exceptions. "Red" and "pig" are W terms, if any are, yet clearly "Red is a color" and "A pig is an animal" are hard to regard as a posteriori, synthetic, and contingent. What this shows, however, is that subsumption of species under genus, without mention of differentiae, is a given in the structure of language. After all, we should not expect language to be without inherent relations between its terms. This structure is far from the least important part of what the child learns. It is a mistake, however, to suppose that the ability to get around in this lattice, to classify terms correctly, necessarily involves any knowledge, even implicit, of concepts in the sense in which that word is here being used.

2. All true propositions, all of whose terms are proper θ terms, are a priori, analytic, and necessary. By "proper θ term" I mean a θ term which is operating within the theory or theories to which it belongs. The qualification is needed because one may make extratheoretical re-

20. Except in identification; see below.

marks using some θ term, for example, that the number of the planets is nine, that the number thirteen is unlucky, or that some cephalopods are radially symmetrical.[21]

3. Propositions in ϕ terms have the same characteristics as those in θ terms: Within the theories to which they belong, they are a priori, analytic, and express necessary truths. (To satisfy the logicians, existential statements must be read as hypotheticals.)

4. W–θ identifications, which produce ϕ terms, constitute a special and unique case. They are attempts to state *what a thing is*. If they succeed in doing so, then they are necessarily true. For a thing cannot be different from what it is. If sulphur is the element of atomic number sixteen, then not only is this true in this world, it is true in every possible world which contains sulphur. For anything not having this atomic number, however like sulphur it might otherwise be, would not be *sulphur*. These identifications are, however, a posteriori. The analytic/synthetic distinction doesn't really apply to them; though, if someone insists, it should be all right to call them "nonanalytic," or even "synthetic" with this mental reservation.

Now to return briefly to the three Venuses. We point the telescope at the bright spot and say "Venus at 2315 GMT, 4 hrs 57' 35" rt ascension, 10° 45' 54" altitude." What kind of statement is this?

It depends on the astronomer. If he is Tycho Brahe with as yet no settled theory, he is talking about Venus-W and making a synthetic, a posteriori, subjectively contingent statement. If he is post-Newtonian he

21. Possibly further restrictions have to be put on to take care of cases arising from Goedel's theorem.

is talking about Venus-ϕ and wasting his time, unless he is calibrating or collimating his instrument or finding out where he is; for the statement is a priori, necessary, and analytic: The theory says already where Venus-θ, hence Venus-ϕ, is; there is no need to look. But what if he is checking the theory? And suppose he looks where the theory says to look, and there isn't anything wrong with his instrument or its collimation, and Venus—W, θ, ϕ, who cares?—just isn't there? This embarrassment is at least imaginable.

He would have a wrong theory on his hands; that's all. "Wrong" when said of a theory just means failure to apply to nature. Something like the contretemps here envisioned has in fact happened three times in the history of astronomy. The first was the failure of Uranus to be where the theory said it should be, which led to the postulation and subsequent observation of Neptune. This was not the rejection of the core of the theory but the adjustment of some numerical values. So also for the similar discovery of Pluto.

It was otherwise with Mercury. If the usual accounts are correct, Mercury was just not on its orbit, and the deviation was not accountable by adjustments of numerical constants in the theory. So the theory, Newtonian mechanics, was thrown away, and another theory, that of Einstein, put in its place. That is, the Newtonian Mercury-ϕ came to an end (*memento mori* when it comes to ϕ terms), the Newtonian θ component departing to be replaced by the Einsteinian opposite number. Thus, with concepts we have a reversed Pythagoreanism: The same body goes on, but with successively different souls.

Now to return to the question that generated this discussion: whether "The morning star is the evening star"

illustrates strict but contingent identity. The upshot is
that the proposition "Hesperus = Phosphorus," though
a posteriori and nonanalytic, is, if true, necessarily true.
To put it as simply as possible, Hesperus is the very
same thing as Phosphorus; and there is no possible
world in which a thing is not the very same thing that it
is. The fact that it had to be discovered using instruments, drawings, and calculations does not alter the
situation. To be sure, if someone began with the *concept*
"Hesperus," that is, somehow had, a priori, a grasp of
of what it was to be Hesperus, Hesperus-ϕ, and similarly for Phosphorus-ϕ, he would have needed no observatory to conclude that they were the same. But that,
as it happens, is not the way we acquire knowledge in
this world.

"Lightning is an electrical discharge" is a simpler kind
of identification than "Hesperus = Phosphorus," being,
so to speak, only half of it: the direct identification of a
W term, lightning, with the θ term electrical discharge,[22]
with the result that lightning becomes a ϕ term. Nevertheless, it may be useful to discuss this example, for it
represents an important class of propositions whose
nature has sometimes been misunderstood.

"Lightning" or an equivalent term occurs in every
natural language, as a label designating a certain kind of
sudden, brilliant illumination of the sky, seen usually as
coming from a jagged incandescent streak and usually
followed by a loud noise. But the fact that we can give
this sort of description must not be taken as implying
that people who use the term correctly could give this or

22. Or perhaps we should say that "electrical discharge" is a ϕ term,
that is, a θ term already identified as referring to a natural phenomenon (discharge of Leyden jar, for example).

any description or that somehow they must have one in the back of their minds. This ordinary W term is a label, a designator, and nothing more. It would be fatally misleading to think of it as signifying a "concept." People are taught, by ostension, to recognize lightning, and they do recognize it, that is, they apply this label appropriately. But there are no formal criteria; there is no check list of conditions that have to be satisfied. The jagged streak may be absent; the noise may not be heard. The question "What is lightning?" occurs to every people, and answers ranging from "Zeus's thunderbolt" to "electrical discharge" have been supplied. The theories specify necessary and sufficient conditions for certain theoretical, postulated entities—anger of Zeus, flow of electrons from region of surplus to region of deficiency relative to protons—to manifest themselves as temporary intense illumination. Again the statements involving only θ terms are, if true, necessarily true; the propositions about lightning-W are contingent (subjectively); and the identification "Lightning-W is lightning-θ" is necessarily true (if true at all) even though knowable only a posteriori. It is the statement that the visible event, hitherto only designated not explained or conceptualized—there is no nontheoretical *concept* of lightning—has those characteristics from which the concept lightning- θ, or (somewhat loosely speaking) electrical discharge, has been constructed.

The identification could go wrong in two ways:

First, if nothing that is correctly labeled by "lightning-W" has, in fact, the properties of electrical discharge. This would be the case if the theory were "false," as ordinarily meant; though we are to remember that the theory itself, barring internal inconsistency, would still consist of necessary propositions, without, however, applicability to the world; which would be a practical drawback.

Second, some lightning-W might be electrical discharge, some other might not be. This would be the case for a tribe whose word which we translate "lightning" is meant to apply also to searchlight beams ("slow straight silent lightning") and nuclear explosions ("mega-lightning").

The second case introduces some instructive complications. When a theory becomes widely known and accepted, its θ terms tend to gain currency to the point of fusing in ordinary speech with the correlative W terms. From an initial condition of having only a W or label term, such as Phosphorus or lightning, together with an esoteric doctrine of what is behind them, we pass to a condition in which the vulgar, having become aware of the constructed concepts, adopt them, more or less, into common discourse. Thus, after everybody has learned that lightning is electrical discharge, water is H_2O, and epilepsy is demonic possession, the W terms of the marketplace merge insensibly into the ϕ condition; people begin to talk ϕ language, like prose. Then the W–θ identification has become "true by definition," or as some would call it, "analytic," to the point that when something ostensibly labeled by the W term turns out not to have the θ properties, it may forthwith be declared "not a real W after all." Thus, the aforementioned tribe might, upon enlightenment, reject searchlights and nuclear blasts as not real lightning (*might*; they would have the option of speaking henceforth of different kinds of lightning), just as we have decided that whales are not really fish.[23]

The identification could *not* go wrong in the following

23. "*fish*, n. (Pop.) animal living in the water, (strictly) vertebrate coldblooded animal having gills throughout life & limbs (if any) modified into fins." *Concise Oxford Dictionary*.

way: While continuing to agree that in this world lightning is electrical discharge, we discover another world
in which lightning is not electrical discharge. There
might be worlds in which something very analogous to
our lightning, perhaps indistinguishable from it to direct inspection by the senses, was, nevertheless, not
electrical discharge but some other kind of phenomenon. But what there cannot be is another world in
which what we now mean by "lightning" is not an electrical discharge. For the ϕ term has taken over in the
language. It can be expunged only by the overthrow of
the whole theory in which it is embedded, not by any
natural catastrophe whatsoever. So if "contingent" is
taken to mean "not true in all possible worlds," we
must deny that "Lightning is an electrical discharge" is
a contingent truth. Hence, this example also fails to
illustrate contingent identity.

Nor, finally, can we discover contingency in "The
gene is the DNA molecule." Here we have an identification not of a W and a θ term but of one ϕ term with
another from another theory. This is part of what is
meant by the reduction of one theory to another. It is
easy to see what to say about this case along the lines of
what has been already observed.

One of many important things Austin said about pigs
is that there is no list of necessary and sufficient conditions for being a pig.[24] We learn to recognize pigs, not

24. J. L. Austin, *Sense and Sensibilia*, (Oxford: Clarendon Press, 1962)
p. 120ff. But Dr. Johnson had anticipated him:
  BOSWELL: "He says plain things in a formal and abstract way,
  to be sure; but his method is good, for, to have clear notions upon
  any subject, we must have recourse to analytic arrangement."
  JOHNSON: "Sir, it is what everybody does, whether they will or
  no. But sometimes things may be made darker by definition. I see
  a *cow*. I define her, *Animal quadrupes ruminans cornutum*. But a goat
  ruminates, and a cow may have no horns. *Cow* is plainer."
  (*Life of Dr. Johnson*, April 7, 1778.)

by checking off their special marks (as we may learn to recognize syllogisms in Darapti, 1968 Volkswagens, or least bitterns), but by their Gestalt, by "ostensive definition." Nor, having learned to recognize them, can we then proceed to list all (and only) the characteristics that an object must have to be properly called a pig. The word "pig," which we use to label certain beasts we come upon in the barnyard and of which these remarks are true, is a paradigm W term.[25] Not only names of things, but names of qualities (red), relations (in), actions (run), and abstractions (cause, probability) may be W terms.

25. Some philosophers hold that "reference without at least some description would be altogether impossible." (John R. Searle, "Proper Names," *The Encyclopedia of Philosophy*, Paul Edwards, ed., Vol. 6, p. 491.) They argue that the only way we can get from language to the world is via descriptions. Therefore, even proper names have some sense, for example, Aristotle = the pupil of Plato and teacher of Alexander the Great.

If this view were correct, then there could be no W terms at all, for if ordinary proper names must represent concepts, then *a fortiori* so must common nouns like "pig."

But the doctrine is nothing more than a prejudice, akin to such familiar contentions as that in order to do something overt, such as to lift one's arm, one must previously do something internal, for example, do a volition to lift the arm; or that I do not know, but only infer, the existence of other minds. The plain fact of the matter is that there is such a thing as ostensive definition: Confronted with the thing, we are informed that a certain word signifies it, and that is the end of the matter—we are neither provided with a description of the thing, nor in any way obliged to provide one for ourselves, not even for internal reference. My total inability to describe the differences between the faces of my two sons or in the tastes of oranges and grapefruit, in no way handicaps me in telling them apart.

The name of an historical personage such as Aristotle may be said to have a "sense." We do not know Aristotle otherwise than as the pupil of Plato, author of the *Nicomachean Ethics*, and so on; and perhaps some descriptions of this sort underlie our references de facto. So "Aristotle" is, for us, a ϕ term, or analogous to one—a concept in a way. However, it does not follow, and indeed could not have been the case, that it was a concept in little Aristotle's parents' and playmates' thinking.

On the other hand, every language contains some terms learned typically by explicit definitions, which are intellectual constructions out of relatively simple notions. Examples are the number seven, uncle, checkmate, promise, sodium, sodium yellow, phlogiston, electron, witch, plaintiff, unicorn. These are fit to be embedded in "analytic" sentences and concatenated theories, for when our talk is restricted to them, we can know exactly what we are talking about and what the logical interrelations are of the statements we make. Their structures are intellectually transparent. There is no more in them than we put in.

The point has often been made that as we learn more and more about things, our concepts change so that what starts out as a synthetic truth ends up as an analytic sentence. We begin by using the word "cloud" merely to label a white or gray fuzzy-looking thing in the sky.[26] What would it be like if we got up close? We don't know. A dense mass of fibrous tissue, perhaps. Later we do get up close, and we find out that it is a mass of water droplets in continuous motion, a fact which we might express in the synthetic proposition "A cloud is a mass of water droplets." But after we have learned this from close-up observation of many clouds, what then? Do we any longer allow the possibility of a cloud's turning out to be a dense mass of fibrous tissue? We do not. Well, but suppose we actually found one—surely it is conceivable that we should? —Found one *what*? Cloud that was a dense mass of fibrous tissue? No. Having gone this far, we would have to say, "It

26. Or more eloquently, "a large semitransparent mass with a fleecy texture suspended in the atmosphere whose shape is subject to continual and kaleidoscopic change." (Place, *op. cit.*) However, this is still not a *concept* in the sense in which I am using the term.

looks very much like a cloud, but it isn't really; instead it's a dense mass of fibrous tissue." "Cloud" and "dense mass of fibrous tissue" have become incompatible.

What has happened is that the term "cloud-W," which is just a label, has fused with the term "cloud-θ," which designates a concept in a meteorological theory, to form "cloud-θ," a term both labeling a cloud and telling what a cloud is—encapsulating a theory. This ϕ term, which has both identifying and theoretical functions, has replaced the W term in ordinary speech. Consequently, to use it is to run a risk, for all statements in which the term is used to identify objects are now subject to error. Perhaps beginning tomorrow many or even all the objects we are wont to label as clouds will turn out to be dense fibrous masses. To go on calling them clouds would then be a mistake. In that case, however, we will not have to take back what we now say about clouds, namely that the concept of a cloud is that of a mass of water droplets in suspension. We shall have found that clouds are extinct, pseudo clouds having usurped the skies. We shall not have moved into another "possible world" in which clouds—real clouds—are dense fibrous masses. For us, now there exists a check list: It is a necessary condition for something to be (correctly called) a cloud, that it be a mass of droplets; a characteristic logically incompatible with being a dense fibrous body. So "Yonder clouds are masses of tiny droplets" is a necessary truth if true at all, that is, if yonder clouds are indeed clouds and not something else mistaken for clouds.

A slightly more sophisticated example may help to bring out the linguistic features to which I am calling attention. Let us consider sulphur. Here again we have no ordinary "concept" at all, no list of necessary or suf-

ficient conditions; we have only a word labeling a stuff
that some of us learn to recognize by its yellow color and
powdery texture (neither of which sulphur always has)
and that we use to dust our grapevines and kill our
dogs' fleas. We do not have much trouble in telling what
is sulphur and what is not. For one thing, most of it
comes in containers clearly marked "sulphur"; for another, there are not so many kinds of yellow powders in
the world. We could be fooled by a plastic imitation. The
cleverer among us might try burning it, sniffing for the
acrid odor. But that test would not be decisive.

Comes the physical chemist with Mendeleyev's chart.
Sulphur is element number sixteen, he informs us. It is
composed of atoms, which are all (almost) alike, having
sixteen protons in the nucleus, with an equal number of
planetary electrons. Now the chemist's statement is
qualitatively different from the veterinarian's or the
vineyardist's. The chemist has a comprehensive theory,
in which the elements bear certain structural relations to
each other so that they can be serially ordered by
number of nuclear protons. That whose name comes
sixteenth on the list has, by virtue of the structural
properties that put it just there, certain characteristics
that account for its ability to kill bugs. The theory says
that if anything is composed of atoms with sixteen protons in the nucleus, it will react with other atoms in certain specified ways. These are hypothetical necessary
propositions within the theory. Moreover, the chemist
goes on to say, that which is assigned atomic number
sixteen in the theory is realized in the world by sulphur:
The θ term "element No. 16" is identified with sulphur-
W. And from now on, when a serious question arises as
to whether a certain stuff is or is not sulphur, we have a
decisive criterion: We determine whether it has atomic

number sixteen. If it has, it is sulphur; otherwise, it is not. And this is final, or, if we prefer to call it so, "analytically true." Sulphur-ϕ has replaced sulphur-W. To disentangle them at this point would be to reject atomic physics altogether.

Most of the changes in the meanings which terms are typically used to convey consist in the explicit incorporation into the standard meaning of characteristics that not only serve to make more exact and sophisticated identification procedures possible, but also are fit to take their logically allotted places in the system of the world. Pure W terms are pointers; with them alone at our disposal we could only describe a blooming buzzing confusion. But clouds as water droplets, rather than as mere large semitransparent masses, have become elements in the science of meteorology. In grasping their nature the ancient Greeks began to see them not as loose and separate but as phases in a great cycle of evaporation and condensation. This understanding has passed into the common consciousness, has become part of everyman's concept of cloud. The dogma of a fixed "ordinary concept of cloud," applicable indifferently to cumulus, nimbus, and angel's bed, is a curiosity that might be stigmatized as linguistic Ptolemaism.

To summarize the conclusions reached thus far:

1. The examples often offered of identity that is both strict and contingent fail to do so because the identities offered are necessary, that is, the conjuncts would not be distinct in any possible world.

2. These identities nevertheless are established a posteriori.

3. In those cases where the effect of a statement of strict identity is to create a concept where before there was only a label, the question whether such a statement is itself analytic or synthetic is an improper question.

Two questions remain: Why is the theory of contingent identity so popular? And if there is no such thing as contingent strict identity, what effect does this fact have on theories of mind and body?

One motive for urging the theory of contingent identity has been the desire of its proponents to assure scientific status for a materialist theory of mind, joined to the belief that all crucial statements in the natural sciences are contingent. The defenders of contingent identity insist that it was an empirical discovery that sensations are brain processes, just like the discovery that clouds are water droplets; and a genuine empirical discovery must always be reported in a contingent statement. Furthermore, if "Sensations are brain processes" were a necessary truth, its denial would amount to a self-contradiction, which obviously it does not.

As we have seen, there need be no fear of forfeiting scientific standing through indulgence in necessary propositions. However, from this nothing follows respecting either the logical or the honorific status of "Sensations are brain processes" and kindred slogans. Let us now turn to consider these matters.

Typically, these sentences are thought of as having words for "mental concepts" on the left hand and for "physical concepts" on the right hand. Before the days of the contingent identity theory, they were also thought of as expressing attempts to "reduce" mental concepts to physical or as providing physical "analyses" of mental concepts. The main innovation of the Australian theorists has been to discard this way of characterizing their function and to say, instead, that they are statements of identity though not of "meaning," thus conceding the main objection of the linguistic nonreductionists and presumably converting them to the substantive materialist doctrine.

All these views, including the contingent identity theory, take for granted that there exists in ordinary language a "vocabulary of mental concepts." The supposition has been accepted at least since the time of Descartes. Nevertheless, it is dubious. Never outside the philosophical closet, and seldom even in it, do we construct locutions with "mind" as subject and as verb one or the other "mental activity" word such as sensing, thinking, knowing, or willing.[27] Nor is "mental feeling" a pleonasm,[28] nor "bodily sensation" a contradiction.[29]

But there is no need to belabor the point, for if a term is to signify a mental concept, it must at least signify a concept; and this the candidates, at least most of the famous ones, fail to do. Words like "sensation," "pain," "thought," and "reminiscence" are like "pig" and prechemical "sulphur"; we know what to apply them to, how to use them as labels, but we have no check lists by which we can decide whether what we are dealing with is a real specimen. We are even worse off with these terms than with "pig"; we know what category to put "pig" in, but we do not know even that about "sensation" and the rest. "The mental" is a pseudo category fashioned to mask our ignorance on this issue. *Pace* Ryle, there is no concept of mind.

What we do have in this curious pigeonhole is a set of W terms of the most primitive sort. On the other hand, "brain process," "central nervous system," "C-fiber stimulation," and the like are rough-and-ready θ terms, or placeholders for the theories which we hope will come some day. So "Sensations are brain processes" is

27. See my paper "Spinoza's Theory of Mind," *The Monist*, 1972, p. 568f.
28. J. L. Austin, *Philosophical Papers* p. 76.
29. D. M. Armstrong, *Bodily Sensations* (London: Routledge & Kegan Paul, 1962).

the beginning—only the beginning— of a W–θ identification.

W–θ identifications are necessarily true if true at all. But that, one has to admit, is not a very plausible thing to say about sensations and brain processes. As the contingency men insist, there is no aura of self-contradiction surrounding "Sensations are not brain processes." We can, so it seems, imagine disembodied existence and thought. One may reply that there is no reason why necessary truths have to be self-evident and that we can imagine all sorts of contradictory things. But these answers do not remove our worries.

The trouble is this: When we ask, "What is a cloud?" and are told that it is a mass of water droplets, we are satisfied that now we know what a cloud *is*. Similarly for lightning, the gene, the morning star. But when we ask, "What is a pain?" and get the answer "A stimulation of C-fibers," even if we understand the theory behind the remark we are still left wondering, "But what *is* a pain?" Some condition for knowledge, perhaps a psychological one, has not been satisfied.

What is lacking, I suggest, is something we might call bilateral symmetry. This quality is most evident in the case of morning and evening stars. We can see, or picture to ourselves, the morning star and the evening star in the same way, and also we can imagine ourselves at a more advantageous location in space where we could directly observe the morning star "becoming" the evening star. We can view a cloud from a distance, and then again close up; we understand that it is the same thing we are looking at from both viewpoints. And though we can see neither genes nor DNA molecules, nor directly count the protons in a sulphur atom, we can form imaginative pictures of what it would be like to do so. In all such cases, we have on the left-hand side of the iden-

tification a W term which is a label for something that
we perceive, or can imagine ourselves as perceiving,
that is, that we can stand to in the relation of spectator.
On the right is a θ term that either designates something
likewise perceptible (water droplets) or a model based
on perceptible objects. Thus, we can grasp in imagination how the right-hand expression denotes the structure of what is labeled by the left-hand expression.

This does not happen when a certain kind of W term
is on the left, namely, when it is a label for something
that we can never be spectators of, even in imagination.
Sensation, pain, and perception are some of these
terms. They signify the ways we get in touch with the
world, but being the gettings-in-touch, we do not get in
touch with them; they are not, for us, objects. But the
right-hand terms signify objects of perception or things
modeled on them. Even if we never peer through the
autocerebroscope, we have no trouble imagining what it
would be like to do so. We would literally see brain
processes. We would see, for example, a twinge, or the
cerebral portion of the process of looking through the
autocerebroscope. But we cannot imaginatively link this
up with the corresponding W term, which, though it
labels the very same thing, does not label it qua object of
perception but qua perception itself.

So in one order of knowing, I know what a pain is: I
recognize it when it occurs and I apply the word to label
it. In another order of knowing, I know also what a pain
is: It is a stimulation of the pain receptors and the consequences thereof in the brain. It is the very same thing,
but the ways of knowing it are not the same. The
linkage between them must remain abstract. We have
indeed an irreducible dualism, though not of objects.[30]

30. Thomas Nagel has made substantially the same point in "What
Is It Like to Be a Bat?" *Philosophical Review*, October 1974, p. 445,
note 11.

It is consequent on the difference in these ways of knowing that to know something in the one way neither entails nor excludes anything about knowing it in the other way. That is why we can imagine disembodied existence, and its converse, implicit in epiphenomenalism and some kinds of behaviorism, of the possibility of outward bodily behavior as we know it without any conscious accompaniment. It is why people continue to be tempted to talk of psychophysical correlations or causal relations instead of identities. It is also the reason why the mind-brain identification is a philosophical thesis, a very general framework assertion about the way in which we can conceptually organize the facts rather than just another fact—even a very important one—to be integrated into a preexisting schema. It is the assertion that the brain physiologist *is* studying the mind, that there is nothing more to do, at least nothing more ultimate than what he does.

III
PRIVACY

In the preceding chapter we sketched in broad strokes what we take to be the content of the slogan "Sensations are brain processes" and saw that this proposition can be intelligibly construed as expressing a necessary truth. But we have not shown that it *is* true.

I do not claim to know whether it is in fact true. So I shall not try to prove it, but only to show that some objections, which if sustained would put it out of the running, are not valid. In this chapter we shall consider the so-called privacy of the mental, which some philosophers hold to be a fact and incompatible with an identity theory of mind and brain.

The story goes this way:

The motions of my muscles are publicly observable and detectable. But I alone really know what my thoughts and feelings are. I may tell you about them or exhibit them in various ways, but I don't have to; I can keep them to myself if I am determined to and have enough self-control. Some feelings are indeed hard to conceal or to feign; but even of murderous rage or dark despair the public sees only the symptoms, never the real thing behind them. The one passion that is impossible of concealment or counterfeit, in the male anyway, is significantly enough called "physical," though even there the distinction between feeling and symptom remains.

Common sense, so the story continues, and the law recognize this distinction as basic. Even tyrants seldom try to punish wrong thoughts as such; the crime of imagining the King's death really consists in talking about it gleefully where others can hear you.

According to this story, then, the things of which we are aware fall into two classes, the public or outer and the private or inner. What more than one person can be directly aware of is public; the private is that of which it is not just difficult or physically impossible but inconceivable for more than one person to be directly aware. Thoughts, feelings, urges, sensations, pains, and the like go into this second bin; everything else goes into the first. Since Descartes, this division has usually been taken as defining the distinction between the physical and the mental.

Since all physical things and events are in principle observable by more than one observer, hence public, the putative existence of a class of private entities seems to negate materialism. Perhaps not quite. Privacy does not strictly entail a substantive dualism, because the private entities might turn out not to be things in their own right but rather aspects or features or other nonsubstantive concomitants of physical entities. Yet the distinction in its common form certainly tends to support a dualistic picture of things. That is why plain men often are metaphysical dualists or can easily be made to believe they are.

Although the distinction is a common one, the terms in which philosophers describe it are not. "Private/public" and "inner/outer" are metaphors. Let us examine them.

"Private" and "public" operate outside philosophy in ways like these.

My private collection of Rembrandts contains pictures some of which were once in other private collections, some in public collections. I can put my private collection on public view if I choose, though it does not thereby become a public collection. If I do so, I may withhold a few specially choice items, which are thus more private than the others. Some I may not let anyone else ever see, though that doesn't make them uninspectable in principle. More than one person at a time—a married couple for example—may jointly own the same private collection. And there is no generic difference between the items that go into private and public collections, nor in the ways in which they are seen or otherwise inspected.

People have public lives and private lives and try with walls, window blinds, watchdogs, and wrought-iron grills to keep what is private from becoming public.

Both public and private property may be open or closed to the public.

We see from such examples that in their literal employments the words "public" and "private" stand for contingent relational properties of things and events. Nothing is by its own very nature either public or private, and anything that is the one can conceivably become the other. The 'public'/'private' distinction of the philosophers, on the other hand—which I shall indicate hereafter by the single quotes—is supposed to mark intrinsic, necessary properties of things: on one side, the things in the external world ("the external world" being a phrase derivative from the other metaphor, of 'inner' and 'outer'); on the other, the thoughts, feelings, perceptions, and the like that inhabit the mind,

or perhaps even compose it. Between the two there is
no traffic.[1] Everything that is, is one or the other for all
time. These essential qualities, furthermore, do not
admit of degree—it makes no sense to divide my mental
images into more and less private ones. Nor can two
persons, however intimate, share anything really 'private.' Finally, although epistemologists who make use
of this distinction do not agree on how or even whether
'public' things are inspected and become known, they
insist that the mode of our knowledge of 'public' things
is very different from the manner in which we become
acquainted with the 'private.' It is not easy to specify a
single respect in which the public/private and 'public'/
'private' distinctions coincide. Even where they both
involve exclusiveness, they do so in quite different
ways.

The ancient 'inner'/'outer' distinction gets some nourishment from idioms of English and other languages
such as "I had it in my mind to . . . ," "I can't get you
out of my mind," "What did you have in mind, Mad-
am?" "He is out of his mind," "It's all in the mind."
However, in these usages speaking of something as
"out" of the mind refers to forgetfulness or madness,
not to objectivity or 'the external world.' 'Outside the
mind' is a philosophical expression, not interchangeable
with "out of the mind," a phrase which indeed does not
occur in colloquial English; nor is 'in the mind' synony-
mous with "in mind." The latter expression means
something like "in the center of attention" and does not
imply or even suggest the mind as a container for 'mental objects.' That seventeenth-century theory has no

1. Save in the metaphysics of neutral monism, which we shall note
below.

recommendation either in the facts or in vulgar prejudice. We may allow that we have thoughts, without
being in any way impelled to provide containers for
them.

So these metaphors seem rather unapt. But do they
do any harm? Well, like all metaphors, they present
models; and when the models are as bizarre as these,
understanding can hardly be furthered. Since a metaphor commonly induces us to ascribe its extraneous features to the phenomenon described (we are surprised if
a man called "a lion" turns out to be bald and high-
voiced), we might expect 'private'/'public' to tempt in
two ways: to the supposition that things and thoughts
are both, so to speak, *things*, that is, individuals of some
kind, since only individuals can be literally possessed;
and to reinforce the tendency to think that thoughts as
well as things are objects of inspection. And we find
these temptations being yielded to in much philosophy.[2]

The hazards of the mind-as-container model have
been recognized. But perhaps the damage wrought by
this and other models (such as the mind as workshop,
where 'operations' are 'performed' on 'data' or 'materials') is subtler and more persistent: we are induced to
keep on assuming that every human being has something called 'his mind,' constituting a unity in its own
right, and with individual components—this thought,

2. Perhaps neutral monism may be explained as an extreme case of
being taken in by this metaphor. That curious doctrine holds that the
fundamental stuff of reality is neither matter nor thought but something in between that can manifest itself in either way, depending on
context. It is not very clear what is meant by the suggestion that a brick
and a thought may differ only in their surroundings; but one might get
this notion as a result of taking the 'private'/'public' distinction to
carry with it from literal employments the feature that one and the
same object may move from one side of the barrier to the other.

that feeling, t'other impulse. Having thoughts is still conceived of as like having possessions, and the thoughts one has are even supposed to be countable as are one's garments, books, and yachts. And if they are no longer held to be all in one big bag, at any rate they must be somewhere in a single neighborhood—perhaps of the brain, where they become candidates for "correlation" with events at the synapses, etc.

Yet the 'public'/'private' contrast strikes a chord. Even if the terms are badly chosen, there is something important, we feel, that they are intended to mark. I can't feel your feelings, think your thoughts, suffer your pains, though I can see your pictures, wear your socks, pay your debts. In some way we are shut up inside ourselves ('in' again); and this being-shut-up is coextensive with what we think of as the life of the mind, our sentience.

Some would dismiss this as an uninteresting fact about the possessive adjectives. I can't play your tennis game, grow your hair, or have your indigestion. So what?

However, this seems too facile a line to take. Consider the myth of Tiresias; it fascinates us because we recognize that a man can never really know what it's like to be a woman, nor vice versa. Or if someone thinks he can, let him try to imagine being a cat, a spider, or a pair of ragged claws scuttling scross the floors of silent seas.

Now if we cannot describe this situation correctly in terms of a division of the world into 'public' and 'private' objects, how are we to describe it? Thus far we have seen that it is doubtful whether the objects we are concerned with can be helpfully described as private. But there is a more fundamental question. Remember-

ing the injunction not to multiply entities beyond necessity,[3] we should ask whether there are any objects at all
in the offing.

The point that the story about 'private' objects is trying to make concerns knowledge. It is plausible to say
that whenever I know something, there is something
that I know. This something is the "object" of my
knowledge. And it is plausible also, in general, to
maintain that there are two ways of knowing it: directly
and indirectly, by acquaintance and by description. The
object of the knowledge that classical scholars have is
ancient Greece, which they know indirectly or by
description. Socrates, on the spot, knew it directly and
by acquaintance. A blind man may learn physics, including optics, and in consequence know everything
there is to be known about color. But his knowledge is
indirect and by description whereas sighted people
know it directly and by acquaintance. Now it is just a
contingent fact that people live at one time or another or
are blind. So it is merely contingent that some people
know ancient Greece or colors directly, others only indirectly. But let us suppose that there were something
which, as a matter of logical necessity, could be known
directly by only one person. What he knew, in that case,
would be a private object.

There are such objects, the theory contends. One's
pains, for example; in general, one's sensations—or the
"immediate objects" of these sensations. The verb
"know" in first person constructions with these objects
signifies this mode of immediate acquaintance; in the
second or third persons, mediate or descriptive knowl-

3. Or as the late Paul Marhenke more accurately put it, the impossibility of doing so.

edge. I know my toothache by acquaintance; the dentist knows it by description.

The theory has many refinements, and exists in different versions, but this much will do for the purpose of inspection.

The verb "know" figures in many grammatical constructions, of which four are particularly interesting: know X; know that . . . ; know how . . . ; and know how to. . . . In only the first of these[4] is there a full-blown, unambiguous object of knowledge (Socrates, Venice, etc.) and not always even then. For I may be said to know violin making. But that would more naturally be put in another form: I know how violins are made, or I know how to make violins. I have the know-how; there is no need to postulate an "object." Still more strained would be the-irrationality-of-the-square-root-of-two as an answer to the question "What is the object of my knowledge if I know that the square root of two is irrational?" Perhaps there still exists a Platonic tendency to think of mathematical knowledge as seeing with the mind's eye an effulgent entity in a place beyond the heavens, but the allegorical nature of this picture should be obvious.

So it is not the case that whenever the verb "know" occurs, there has to be straightforwardly some entity that is the object of knowledge. The question that confronts us can be put "To what extent is knowing hunger or heliotrope or femininity similar to knowing Socrates or Venice, or even the binomial theorem?"

Put this way, the question almost answers itself. "I

4. Marked in many languages by a separate verb from the one used for the rest: *kennen* as opposed to *wissen*, *connaître* instead of *savoir*. But this is not exactly the distinction, nor is the German distinction precisely equivalent to the French.

Privacy

know hunger" can be paraphrased, "I know what it is like to suffer from hunger"; but "I know Socrates" does not mean, "I know what it is like to be Socrates." Socrates is out there, whether I know him or not; but hunger has no existence apart from hungry animals. Heliotrope shares this objectivity with Socrates (so I at any rate would maintain), but then there is no difference between "I know heliotrope" and "He knows heliotrope"—if either expression is admissible. And Tiresias's knowing what it is like to be a woman is clearly knowing a state, a condition, and not anything that could with plausibility be called an object.

Objects are objective. Where we can objectively speak of an object of knowledge in a locution in which the word "object" is more than a merely grammatical term, the object must be something, like Venice or the binomial theorem, that is independent of its being known. But this requirement at once rules out the possibility of 'private' objects since their partisans agree that they do not and necessarily cannot exist outside awareness.

Thus the theory of 'private' objects of knowledge is an epistemological monster. It depends on the contemplative epistemology, which assimilates all knowing to gazing at an object. At the same time, it is incompatible with that theory, which (for example, in Platonism) in order to be internally consistent must stress the independence of the object from its knower.

This much by way of clearing the field of confusing terminology. The real problem may yet remain. Surely, someone will say, no matter how we talk of it, the fact remains that some things can't be known otherwise than through direct personal experience. You can't know what a mango tastes like unless you taste a mango. I, who am eating one in front of you, can't tell you. Perhaps we should concede the point (though I might say something more or less getting it across—"It tastes rather like a peach in varnish sauce"). But so what? If you have never seen an aardvark, my description of the beast may not be entirely satisfactory to you. It does not follow that there is anything subjective or 'inner' about either aardvarks or experiences of them.

Again, though, such a retort may be thought to miss the point. If the person you are talking to has had the experience in question, you can remind him of it; if not, you may refer to one somewhat like it that he has had; if that fails, as sometimes it will, there is nothing to be done.

All this may be admitted; but what conclusions are we supposed to draw? Granted that words can only evoke feelings, never create them *ex nihilo*, what is it that I have that you cannot have? This we have not located. It is one thing to be lectured to about the chemical reaction of protein to intense heat and another to be burned, and neither is a substitute for the other. But if you tell me you have a burning sensation in your eyes, and I have been burned, I know what you mean, and I know how you feel. What is missing? The element of strictly personal knowledge seems to vanish just when we have it in our sight; or we may say that it is turning itself into triviality after all, like the fact that I can't suffer your indigestion.

Yet surely it is true and not trivial that I can never know, in the full sense, just what it is like to be a cat. And it is not merely because cats cannot speak to me and tell me. We feel that to know, really, what it's like to be a cat, one would have to be a cat. So there is something that the cat knows that we don't and can't know.

Let us go more slowly. There are occupations and

activities which I suppose I could take up, though I haven't and never will.[5] I have never been an undertaker, but perhaps I have the ability to become one. So if I want to know what it's like to be an undertaker, there are two things I can do. I can ask undertakers, "What is it like to be around dead people a lot?" and they may tell me. Or I could become an undertaker, when, ex officio, I would know what it was like to be one. Now probably if I take the second way, I will end up knowing better what it's like to be an undertaker than if I only asked; but is there any necessity in it? Is it not conceivable that in the end I should admit that my direct experience merely confirmed what I had learned from description without going beyond it? I see no reason to deny this possibility. Writers try to tell us what it is like to be a Greek hero before Troy or a silly provincial French woman. Is it logically impossible for them to succeed? Some artists go further and try, perhaps not entirely seriously, to communicate the experiences of beetles and cockroaches. Granted that we have no criteria for success in such endeavors, why don't we? It is true, but not enough, to say: Because we can't be beetles. For we want to know why and to what extent this is a bar to imaginative knowledge.

Let us go at the problem by yet smaller steps. I think I can imagine with fair accuracy what it would be like to be a millionaire or an undertaker because to become one would involve mainly a change in my external circumstances, followed no doubt by a modification of many attitudes and desires. It is harder to imagine being a medieval peasant because from birth his whole outlook

5. This point and its implications were suggested to me by Charles Jarrett. See now also Thomas Nagel's fascinating paper, "What Is It Like to Be a Bat?" *Philosophical Review*, October 1974.

would be different from mine. But the more learned one is in medieval studies, the better approximation one ought to be able to make. Suppose now I try to conceive what it must feel like to be a cat—have I moved across a line from difficulty to impossibility? Though the millionaire, the peasant, and I have to cope with different worlds in different ways, we share a structure and the kinds of abilities that go with it. But cats are built differently and can do things that I can't do—purr, land on their feet, arch their backs, enjoy eating raw birds—and can't do things that I can do—chew their food and, above all, talk.[6] It is this subtraction that seems to create the greatest imaginative difficulty. It is my nature to talk, to others and to myself, and what it would be like to have the sense experiences I have but not be able to put them into words is something that baffles my utmost imaginative endeavors. To say "Everything would be just as it is, except that you couldn't verbalize" is correct but unhelpful. We have, I believe, crossed a great divide at this point. The rest of the way, down to the cockroach and the pair of ragged claws, seems to present no such catastrophic descent.

But we hardly need argument or example to convince ourselves that if some creature is just like us in structure and functions, we have no difficulty in imagining what it would be like to be that creature; if it is totally unlike us, we can't imagine this at all; and in between the difficulty varies with the degree of structural and functional difference. This is a truism; but we can derive from it a conclusion that is perhaps not so banal. It is that our difficulty in imagining what it would be like to be such

6. Wittgenstein's remark that if lions could talk, we wouldn't be able to understand them, makes a point allied to, but not identical with, the one I am presenting.

and such a creature is simply a difficulty in imagining what it would be like to be able to do such and such things. We do not have two difficulties, one of imagining what it would be like to possess certain abilities, and another separate difficulty of imagining what the "mental accompaniment" of that ability might be; we have only a single difficulty. We do not wonder first what it is like to do, habitually and professionally, the things that an undertaker does, and then go on to wonder, in a separate session, how it would feel to do them; the first wondering includes, or simply is, the same as the second.[7] Or to come at it from the other side, suppose that cats behaved as in fairy tales, that is, dressed, spoke, and lived just as we do, except for being able to purr and swish their tails. We would then find the consciousness of cats as unproblematic as that of our friends who have some little ability that we lack, say to wiggle their ears.

To imagine what it's like to be a cat is precisely to imagine oneself being able to do all those things and, what is more difficult, only those that a cat can do. To know, in a strong sense of "know," what it's like to be a cat is impossible for no other reason than that it would require one to have the structure and functioning of a cat, to *be* a cat. So our inability to fathom the 'private' or 'inner' life of a cat is no different from our inability to participate in its outer and public life. It differs only in degree from inability to know what it is like to be a champion boxer. Not to be able to know these things is not to have the abilities, the developed structures in

---

7. To be sure, wondering how it feels *to* chop off heads and wondering how the headsman feels *about* chopping of heads are separate and distinct wonderings. But the attitude that a creature has toward what it does is separate from, though connected with, its experience of doing it.

question. There is nothing hidden, secret, 'private,' or otherwise metaphysically mysterious lurking about; it's just the prosaic limitation that one can't be everything at once.

Now let us review the case of the congenitally blind man. Often it is argued that since he does not and cannot know some things that we know, but can know all the public facts, there must be some 'private' entities, viz., those that we know and he doesn't. But what the blind man lacks is the ability to find out certain things in the way that we, who have sight, can find them out. We can tell that there is a tomato there by looking; he, only by feeling, sniffing, or being told. However, it is a complete non sequitur to manufacture from this misfortune any 'private' object; the only object in question is the tomato, which is as 'public' as anything can be. To know, in this context, is to exercise a certain ability. It makes no sense to talk of a private ability.

Is it behaviorism to say that knowing is the exercise of ability? That depends. "Crude" behaviorism wants to say that discriminating between red and green lights just is stopping one's car for red and going on green. This is not what I mean to advocate. You go on green because you are aware that the light is green. One is tempted here to suppose that the behavior, stepping on the gas, is the effect of a cause which is the "mental" event of recognizing the color, "having it in mind." But that isn't right either. Recognition comes before overt action, but recognition itself is an activity, the exercise of an ability. Discriminating is doing, not just sitting back and contemplating. The event of recognizing the color cannot be the cause, in the billiard ball sense, of stepping on the gas, for if it were, the sequence would be a mere reflex. And it would be hasty to assume that the two events have to be links in the same causal chain.

Thinking is not the motion of the vocal muscles, not even the tiny, barely perceptible motion of them, not even, *pace* Hobbes, the endeavor to move them. Thinking precedes such motion and endeavor. It does not follow that the thinking is not itself an activity, even a motion, though not of the muscles. To say that thought is (gross) behavior is as absurd as to say that legislation is policing or that cookery is eating.[8]

This *is* to deny that there is any thing, or indeed activity, to which I alone have access. It is not to deny self-awareness, that is, that I am usually in a better position to monitor and describe my activities than you are and that sometimes, when I am not grossly moving, I may be the only one who can. But expertness and instrumentation could make up the handicap.

Perhaps some of these contentions will become clearer if we discuss that famous philosophical instrument the cerebroscope.

There is a deplorable tendency to suppose that in the employment of this device, the subject is aware of his sensations 'from the inside' while the observer sees the subject's mental activity 'from the outside.' This danger is not eliminated entirely by the realization that the subject is not observing his own sensations but having them. One must describe very carefully what is going on.

The subject S looks at a tomato T. The observer O looks at appropriate regions of S's brain. What O observes is brain activity A_S. As a consequence of this

---

8. Another, earlier, slogan, "The brain secretes thought as the liver secretes bile," is likewise absurd—like "The muscles secrete work" or "The fingernails secrete growth." Yet it was a mistake in the right direction.

FIGURE 1.

attention on the part of O, there occurs in O's brain some activity A_O.

A_S will not resemble T. However, this in no way creates an objection to the identity theory. For that theory does not say "A_S is what is seen when T is looked at"; it says "A_S is the seeing of T by S." There is no reason why the seeing of A_S (viz., A_O) should resemble either A_S or T. Nor is the situation changed when S = O so that the cerebroscope is modified into Professor Herbert Feigl's ingenious autocerebroscope.

Why has this innocent instrument been thought to present an obstacle to the identity theory? Because of the problem of its collimation. We can only tell when we are observing "the appropriate regions of S's brain" by establishing, empirically, a correlation between brain activity (A_S) and S's report of what he senses. But, so the objection continues, if seeing T just *is* A_S, or even necessarily is A_S, how can this empirical correlation be required, or even allowed? By Leibniz's law, if $a = b$ then every property of a is a property of b and conversely. Now "seeing the tomato" is, on the account here presented, (a) a happening which S reports, and also (b) a happening (A_S) which O reports. But so far from having every property in common, they have none, and they are only empirically correlated. For example, S reports the tomato as gone from his visual field at the same time that O reports the cessation of the activity A_S.

The reply to the objection is simple. While seeing the tomato is the same as A_S, S and O do not both report it, but different things. S reports seeing the tomato; O reports seeing certain brain activity. A_S, which allegedly is the seeing of the tomato. That is, O reports seeing the seeing of the tomato. Watching a voyeur is not being a voyeur.[9] The tomato is what is seen by S. A_S is the seeing of the tomato by S, also what is seen by O. The empirical correlation is not between tomato and tomato, nor between seeing of tomato and seeing of tomato, but between seeing of tomato and seeing of seeing of tomato. There is no reason to expect these to be resembling.

I shall conclude this chapter with some speculations concerning the perfected cerebroscope and its uses.

The brain is not organized in what an electrical engineer would regard as a straightforward manner, with discrete circuits for discrete functions. It is, as they say, wet, accomplishing its tasks in a manner not fully understood, but involving more than mere summations of elementary processes. Thus, although the man who knows the design of a computer can in principle read off from a flow chart the computation in which it is engaged, it is doubtful whether any analogous translation procedure for, say, neuron firings is possible. Nevertheless, we are forced to assume that there is some structural correspondence between brain state and sentient conditions.[10]

9. Save *per accidens*, as in the autocerebroscope.

10. This may sound like a weakening to the point of abandonment of the main thesis of this book, that sentience is just the functioning of the brain. But it is not. I mean that we can and must assume the possibility of discriminating, by vision or instrumentation, between different configurations of the brain in such a manner that we can recognize brain state B as being the sentient condition S. All that is being said is that the physical structure of sentience really exists. Nor is this to suggest that anything of this sort could ever be a practical possibility. In philosophizing about the mind, we must occasionally talk about such things as brain transplants, which must be regarded as in principle possible, though physically impossible on account of the multitude of nerve splices required, as contrasted with the simple plumbing connections of the heart transplant.

So now let us imagine the cerebroscope developed so that these structural correspondences could be read out on a display. When a red light flashed, we would conclude "He's angry"; when a certain pointer gave a reading of 87, we would say "His pleasure has reached the 87 bentham level"; and the ringing of a bell would indicate "He's on the point of discovering the proof of Fermat's Last Theorem."

Now let us ask: Could a cerebroscope report ever take precedence over a sincere report of introspection?

The usual answer is emphatically negative. Reports of introspection, such as "I am in pain," "Red here now," are supposed to be immune from correction when sincerely made and the tongue does not slip. Some philosophers profess to find unintelligible the retort "No, you aren't" to a man who says he is in pain.

However, this view (which is advocated for certain epistemological purposes foreign to the present inquiry) assumes that nothing can go wrong between the sensation itself and the report of it—nothing except willful mendacity or easily recognized tongue slippage. The epistemologist makes these assumptions because he needs "foundations" of certainty for knowledge. On even casual scrutiny, however, they will be seen to be rather implausible. It is a long neurological way from the stimulation of the efferent nerves to the energizing of the tongue and larynx to produce an English sentence, however laconic; and much may go wrong along the way. So it is not a question of the cerebroscopist's report, which is merely empirical, competing against a deliverance of introspection with a logical guarantee; it is just one causal chain clanking against another.

Let us remind ourselves what is necessary and what is empirical in this situation. What is necessary, if the contentions of the preceding chapter are granted, is that a particular instance of brain or neural activity, A_i, is identical with a particular instance of sentience, S_i. What is only empirically established is that the observer O, examining the neural structures, sees or otherwise perceives A_i, that is, does not make a mistake in his observing; or, if the "observer" is an automaton, that the display board has the proper input and its inner parts are functioning properly. So A_i could always be misidentified by O, whether man or machine.

But things are not so simple for the introspector. S has to make a report, verbally, in writing, or by some other signal, and this involves further activity. There is not just the activity of sensing the tomato; there is the further activity of (so to speak) patching the sensing network into the annunciator. Further, there is the innervation of the tongue or finger muscles, etc. Even if in fact nothing is likely to go wrong here, especially in a simple case like recognizing a red patch in strong white light, or the pain of having a fingernail pulled out, still there is a chink in the logical armor. In more interesting cases, introspection is notoriously liable to error.

So the question boils down to that of comparing the relative reliabilities of cerebroscope and protocol sentence. There is no logical barrier to a doctor's or judge's or policeman's preference of the former to the latter.

But if 'privacy' is denied, does the denier not fall into epiphenomenalism or worse? For if everything that can be learned from sentience can be got in some other way, what is the advantage of being sentient?

This question involves a misunderstanding. All the facts about the world, including those about the observer, that one observer can get can also be got by another observer, at least in principle. This is not to say that there are any facts that can be got without observation at all.

The common feature of theories of privacy, the inner, knowledge-by-acquaintance, and the like, as of much in philosophy, is their solving the problem—in this case, how we see a tomato—by duplication: We see, that is, we are acquainted with, an internal unmaterial tomato. Thanks chiefly to Wittgenstein and Ryle this ploy has declined in popularity.

IV

MACHINES

Man is a machine, La Mettrie said[1] a quarter of a millennium ago; and though subsequently he wrote a tract saying that man is more than a machine, the first pronouncement has stuck. Descartes had allowed that the brutes are machines, and man too as far as his body was concerned. Hobbes went all the way:

> For what is the *heart*, but a *spring*; and the *nerves*, but so many *strings*; and the *joints*, but so many *wheels*, giving motion to the whole body, such as was intended by the artificer?[2]

Many people believe that "Man is a machine" is a prime article in the materialist credo. This is a mistake. The slogan, by itself, means hardly anything.[3] It can be taken to mean that man is a structure of interacting material parts. This is true but trivial. Or, at the other end, it has been construed to mean that man is an advanced digital computer feeding into integrated servo mechanisms, which is interesting but false.

---

1. Not exactly: the French word *machine* is a *faux ami* that often, as here, should be translated "mechanism" or "apparatus."
2. *Leviathan*, Introduction.
3. Which is no disqualification for being part of a credo. But there is no materialist credo.

Man is not a machine. Nor is he a mechanism, except in the trivial sense just mentioned. Nevertheless, the analogy of man to various machines, especially computers, is a subject the discussion of which can contribute to a better understanding of what man is, and particularly of what sentience is and why it exists. How much of human behavior can be duplicated, in fact or at least in principle, by insentient mechanisms? Are there any reasons for denying the possibility of a sentient machine? If there are none, what kind of behavior from a machine would show that it possessed sentience?

First, we need to get clear what we are to understand by "machine"—or rather, "mechanism"—and that is not so easy, for these words are W terms. We know what to apply them to, but we have no explicit criteria for their application. If the questions just raised are meant to apply only to machines now in being, the answers are easily given. They are: Very little, if any, distinctively human or animal behavior can be duplicated or plausibly simulated by existing machines. No existing machine is sentient, and nothing that any of them could do would go the slightest way toward indicating sentience. But these facts are of no interest. We want to talk about possible, conceivable machines, for which purpose we need criteria for what would count as a machine.

Examining a quantity of the things we call mechanisms, with a view toward discovering wherefore we call them so, we notice that they are all artifacts—expressions such as "the mechanism of the inner ear" are metaphorical or analogical; they all *do* something, they have definite *outputs*; and what they do is determined at least in part by what is done to them, that is, they have definite ranges of *inputs*. The typewriter on which I am now writing has, as output, black marks on paper; as input, finger pressure on the keys; and for each different key there is a different mark. This one-one input-output correlation does not occur in all mechanisms. One and the same bomb may have but one output for inputs as diverse as being stepped on, being picked up, receiving a certain radio signal, and having a metallic object come within three feet of it. And a Nevada slot machine may come up with any one of a thousand possible outputs for the same input, one dime. Nevertheless, there is a definite causal link between input and output.

Thus, to characterize machines we need to say what an artifact is. And this seems simple enough. Our offhand conception of an artifact is of something that we made, we did not find lying about; and to make something is to take pieces of materials of one sort or another and arrange them in accordance with a preformed plan.[4] These conditions seem to be necessary ones—it is doubtful whether we would call anything a mechanism that did not satisfy them. But they are not sufficient: Babies are not artifacts, though they are consequences of human arrangements of materials in accordance with plans. So are (cultivated) carrots. We might add the condition "not sexually or vegetatively generated," but that is not enough to rule out the *Brave New World* babies

4. The plan need not be definite, and it may amount to no more than a resolve to respond in more or less definite ways to random events. Aleatory music exists, alas, and its instances cannot be denied to be artifacts.

produced in vitro. We can stop up that hole by adding "and not grown." Then an artifact for us is something constructed by us out of materials in accordance with a preformed plan, not sexually or vegetatively generated and not grown. And a machine is an artifact designed to have an input and an output. This definition should be adequate for our purposes, even if counterexamples can be found.

Leaving in abeyance the question whether we are constructed in accordance with a preformed plan, two differences between us and machines are then that we are sexually generated and we have grown. These are big differences. Aristotle thought that if houses grew, they would develop along the very same lines as those according to which the carpenter proceeds: foundation first, then walls, then roof.[5] But he was wrong. If a house grew, it would begin as some tiny thing, faintly resembling a kitchen sink, in the midst of a vast agitated flow of all kinds of nails, sticks, bricks, pipes, wires, pieces of glass, bits of linoleum, doorbells, screws, and roof tiles. Out of this the little sink would somehow attract to itself enough bits of the right kinds of things to increase its own size while retaining the general shape of a kitchen sink. Presently, it would pull apart into two kitchen sinks; this process would be repeated several times until there was a whole agglomeration of kitchen sinks. The outermost sinklets would then begin adding to themselves what was required to modify them into cabinets, ranges, and refrigerators, and so on until the whole aggregate was a structure about six inches each way, beginning to resemble a house. This would con-

5. *Physics* II, 199ª 12.

tinue to attract more materials from the flowing chaos until the house was all there, full size, furnished and with the utilities connected.[6]

Notwithstanding this great gulf fixed genetically between men—or rather, living organisms—and machines, the question whether it is in principle possible to build a sentient machine seems easily answerable in the affirmative. For no matter how the organism got to be what it is, *what* it *is* is an assemblage of material parts. We can, therefore, conceive of its being duplicated, tissue by tissue or even atom by atom. We should end up with what some have called a "meat machine." No such technological tour de force will ever be brought off, but limits on refinement and rapidity of microanatomical techniques, not logic, are the bar to the project. And by definition, as it were, an *exact* replica of a sentient being would be a sentient being. To call this in question would be so far to allow the possibility of a substantive dualism. For if the exact replica monster was not conscious, that could be due only to its lack of a concomitant nonmaterial thinking substance.

Let us assign this hypothesis the name and status of Frankenstein Axiom:

An exact physical replica, however produced, of a sentient being would be itself a sentient being.

This axiom is weaker than the identity theory of mind and body. The identity theory entails the axiom but not vice versa since the axiom is compatible with epiphenomenalism. Thus, every reason for accepting the iden-

6. This propensity of living organisms for selecting what they need from the relatively undifferentiated matter surrounding them is not without its significant parallel in the life of the mind. See Chapter VI.

tity theory is also a reason for accepting the axiom, and there should be others as well—in general, all the considerations militating against substantive dualism. At any rate, I shall assume its truth henceforth.[7]

Later we may find uses for this axiom. In the present context, however, it seems unhelpful. For an answer to the question whether a machine might be sentient ought to aid us in understanding the nature of sentience. But the mere assurance that a replica of a sentient being would be sentient does not advance this understanding. It is possible for us to know the exact structure of every cell, every molecule, in the human body to the point of being able to synthetize it without being any nearer than we are now to comprehending the nature of sentience. To say this is to restate Leibniz's point about walking through a thinking machine as through a mill, finding only "pieces which push against one another, but never anything by which to explain a perception."[8] As everyone knows who has built a hi fi amplifier from a kit, success in putting the thing together is no guarantee of understanding the principles of electricity and magnetism. That grasp is demonstrated only when the builder is able to design his own

7. It follows of course from the identity theory that this axiom is necessarily true; therefore, if it is false, the identity theory is destroyed. And philosophers from Descartes to R. Kirk (see "Sentience and Behavior," *Mind*, 1974) have argued for its falsity. All their arguments depend on the conceivability of a body's having the exact function and structure of one's own and yet being insentient. But reasons offered in favor of this possibility must presuppose the tenability of the 'public'/'private' distinction; otherwise, nothing is presented but the merely autobiographical notice that the writer finds that he can imagine an insentient replica of himself, which is no more to the point than a "proof" that magnetism must be nonphysical because I can conceive something looking, feeling, etc., just like a magnet but not attracting nails.

8. *Monadology*, Section 17.

apparatus, with different layout and components, yet functioning satisfactorily. The mere slave of the Heath-kit manual wanders though the amplifier as through a mill, without ever coming upon anything by which to explain a reception.

On the evidence now available, sentience is not found apart from protein. This suggests, if it does not imply, that anyone setting out to construct a sentient device had better use amino acids as his basic materials, just as makers of abrasives do well to concentrate on carbon and silicon and eschew gold and sulphur. There is no need to suppose, however, that once the principles of sentience were understood, the form of the device would have to be anthropoid—that our meat machine would have to be made in man's image. True, we do not envisage "machinery" made of sticky, slimy stuffs, and we might be disgusted or horrified to see it installed in an office. But that is neither here nor there.

In trying to decide what is to count as a machine in the context of man-machine analogies, we have reached no definite result; nor does it seem possible to draw a sharp line between the natural and the artificial (or synthetic or imitation). Our preconceptions that machines are assembled from a relatively few, mostly rather hard and dry materials must be set aside.

Leaving to our intuitions the decision as to whether some strange object, with a strange history, is or isn't a machine, let us ask how or whether a machine might persuade us that it was sentient. Now the engineer working in the field of artificial intelligence takes no professional interest in this question. If it behaves intelligently, that is enough. But in our inquiry we want above all to know whether an artifact might really think. As we could have nothing to go on but the machine's

externally observed behavior, we would be confronted
with the problem of other minds, transformed from an
academic speculation into a question that would be, in a
queer sense, practical. In this chapter we shall not come
to grips with this problem or nest of problems but
merely skirt its edges.

Sometimes we are convinced that human beings are
still thinking even when little or no overt activity is
going on, for example, in general paralysis. But we
would not extend this trust to a machine. It would have
to convince us of its sentience by quite spectacularly
brilliant behavior. Of what kind and how much? Before
trying to answer, let us consider an argument intended
to show that there never could be convincing reason for
conceding sentience to a machine.

Some people take it as obvious[9] that no machine can
ever be sentient. Perhaps this is obvious with respect to
the machines with which we are actually familiar, but
there seems no reason to extend that verdict a priori to
the soft machines of the future. When argument is
offered on this point, it is usually based on the premise
that a machine "by definition" can do only what it is
designed to do and that this limitation is incompatible
with sentience. There is something in this: sentience, as
we shall see, is superfluous where the only activity in
question is the carrying out of explicit instructions;
therefore, there is no reason to suppose that a machine—or organism, for that matter—that does nothing
else is sentient. Nevertheless, the argument from the

9. Or even as necessary. "But a machine surely cannot think! —Is
that an empirical statement? No. We only say of a human being and
what is like one that it thinks. . . . " Wittgenstein, *Philosophical Investigations*, 360.

premise "*X* is designed" to the conclusion "*X* can't be sentient" is fallacious, as Gunderson has shown.[10] There is no logical incompatibility between being designed and behaving in a way that requires or at least admits of sentience. After all, we might have designed the thing precisely to *be* sentient.

If there is no argument against the possibility of machine sentience, we can proceed to the question of what behavior would be proof of its possession. Now it is possible and indeed easy to state a de facto sufficient condition: The machine does *everything* we do. If a cat conversed with us and showed an average human repertoire of interests and aversions so that its actions were explainable by reference to motives that we assign to ourselves and others, the mere fact that she had the shape of a cat would not cause us to withhold belief that she thought the same way we do. If this would be true of a cat, it would be true of a robot too.

But an objection can be brought against this conclusion, one which takes us back to the argument against the possibility of mechanical sentience. We would allow Puss in Boots to have a human-type mind because cats *to begin with* are at any rate animals; and we already ascribe to them much of the spontaneity, motivated behavior, and so on that we are conscious of in ourselves. We do not design cats; if Puss began to recite Hamlet's soliloquy, it would not be because we had made him so that when triggered that would be his response. But with the robot it's different. We would have designed it. No matter, then, what it did, we could always say "It's acting that way because its machine

10. Keith Gunderson, *Mentality and Machines* (New York: Doubleday & Co., Inc., 1971).

table specifies that output for this input." Consciousness then would be superfluous as an explanation of the observed behavior.

This objection stems from the mistake of thinking of consciousness as something that explains behavior causally. Inferences from overt behavior to the possession of consciousness are not inferences from effect to cause. They cannot be, because consciousness as such never *does* anything. This point will be argued for later; but it should be pretty obvious. However, it does not turn the main thrust of the objection, which is that the robot behavior, being accounted for satisfactorily in terms of programming, could be thought of with perfect consistency as nothing but motion of parts. The reply is that a robot programmed to behave in all respects as we do would necessarily be programmed to be conscious. This I hope to show in the next chapter.

Another approach to this question, which we shall now pursue, is to inquire into what limitations if any there would be on the possible behavior of an artifact that was not conscious. Straight off we can assert a large number of a priori propositions of the form "A nonsentient being could not _____ ." Among verbs that will go into the blank are:

admire	embezzle	long	swear
adore	enjoy	love	swindle
apologize	expiate	mourn	trust
believe	exult	pray	understand
care	fear	pretend	venerate
covet	forgive	promise	wish
crave	gloat	protest	wonder
curse	hate	regret	worry
despair	hope	rejoice	worship
detest	intend	relish	yearn
dislike	loathe	sin	

Some of these actions are impossible for nonsentient beings because the words signify precisely being in a certain state of mind, for example, adore, loathe, mourn, worry, yearn. But not all. Apologizing and promising do not entail the subject's having any particular thoughts. An apology is made by saying "I apologize," a promise by saying "I promise," in appropriate circumstances. But as we know only too well, a computer may be offensive. Why then can it not apologize? Well, apologies must admit of the distinction sincere/insincere. But a nonsentient being could be neither sincere nor insincere. The case of promising, however, is different. One might argue on behalf of the computer that if today it prints out A CHECK WILL BE ISSUED IN THIRTY DAYS and has been programmed to issue that check after thirty days have elapsed, it has satisfied the conditions for making a sincere promise. A sincere promise is simply one uttered with an intention to keep it; and intentions, whatever they are, are not feelings. However, promising is not just stating an intention, even while having the intention; it is also, and separately, the putting of the utterer under an obligation. Insentient beings cannot have obligations, for obligations can apply only to individuals who have their places within some institutional network, involving attitudes, hence feelings, of the members toward one another.

What about understanding and believing? Epistemic verbs are employed by people around computers, who speak of the machines knowing or remembering the facts they print out, and perhaps some of these technicians take themselves to be using the words literally. If they think so, then they are doing so, for these words are W terms that have not yet subordinated themselves to the linguistic regimentation of a theory, despite the efforts of epistemologists.[11]

It is doubtful whether anyone would want to say that a nonsentient being could understand anything. But understanding is a notion of which we will have more to say later.

"Pretend" offers special problems. It is plausible to assume that an unconscious robot might simulate any action whatsoever; could it not then pretend to detest, sin, worry, or whatever?

There seem to be two distinguishable senses of "pretend." One is that of making believe, of let's pretend games, social or solitary. This clearly presupposes sentience. The other, relevant to the deceptive robot, is that of behaving in a manner characteristic of something else, with intent to deceive. So the question reduces to

11. Knowing, for example, does not entail believing, no matter what the treatises may say. I know how to cook an omelet, but it makes no sense to say that I believe how to cook an omelet. It does no good to try to get around this by saying that entailment holds between propositions, not single words or phrases, and that here the proper "analysis" is to the effect that "I know how to cook an omelet" entails "I believe that so-and-so is the proper method for cooking an omelet." For I may never have given the matter any thought, and even if I did, I still might not be able to fill in the directions for which "so-and-so" is a stand-in. To say then that I must do so unconsciously, is merely to invent a "sense" of "believe" for the express purpose of being entailed by "know."

Even with knowing-that (which is not a special "sense" of "know") there is no entailment of belief, only, so to speak, a preemption of the field. If I know that so-and-so is the case, the question whether I believe it can be dismissed as moot. To ask someone whether he believes that . . . is to suggest that he does not know that . . . , and even, usually, to insinuate that . . . isn't true. Believing a proposition is taking an affirmative stance on the question of its putative truth—the word "believe" is cognate with "love." But knowing does not necessarily involve any stance at all. Hence the plausibility of behavioristic accounts of knowledge, and the fact that "I know Gerald Ford is President, but I still can't believe it" is not self-contradictory. "I can't believe it" means roughly "It doesn't fit into my *Weltanschauung*."

whether insentient beings can have intentions. Apparently they cannot. The law, at any rate, takes "intentionally" to apply only to conscious behavior, some psychiatrists to the contrary notwithstanding. Yet what are we to say of a robot that spills ink on the carpet not on account of a malfunction but because it was programmed to do so? Did it not spill the ink intentionally? Maybe this is metaphorical, and we really mean to ascribe the wicked intention to the programmer.

These difficult cases aside, the verbs on the sentience-presupposing list fall into two classes. One, comprising

embezzle	promise	swear
expiate	protest	swindle
forgive	sin	worship
pray		

may be called social, as they concern transactions between persons.[12] The remaining thirty-three, from "admire" to "yearn," signify affective attitudes. The list is of verbs that strike me as *clearly* requiring consciousness. I did not put "will" on the list because I am not sure of the conditions for its proper employment, other than as an auxiliary verb, and *a fortiori* I am bewildered about its relation to consciousness. The more intellectual verbs do not seem clearly to require consciousness.

Consider the list below (the purpose of the asterisks will be explained later).

*acknowledge	*calculate	*confirm	duplicate
*add	*check	control	*enumerate
*analyze	*collate	*count	*err
*answer	*compare	*deduce	*figure
*argue	compile	*diagnose	find

---

12. Actual, putative (for example, in order to pray we must assume the existence of a god and his ability to hear us), or feigned (as when I forgive my cat for scratching the furniture).

index	print	*review	*solve
*know	*prove	sample	teach
*learn	*reason	*see	tend
*mistake	register	seek	test
modify	regulate	select	transcribe
number	reject	sense	*translate
*obey	*remember	*signal	try
order (put in)	repeat	*signify	verify
*plan	reproduce	simulate	write
*predict	respond		

These verbs have all been applied to the operation of admittedly unconscious machines, and some of the people who have used them no doubt supposed that they were using them literally. Most if not all of the "intellectual operations" are included. Clearly in some cases these usages are literally correct: control, index, register, transcribe. Of some others, I for one would say that they apply to machines only metaphorically: It is not even literally true that a machine can count or add—it is a tool that human beings use in counting and adding. For what is counting? It is not enough to answer "bringing into one-to-one correspondence with a base series," for on. this definition the maidenhair fern counts its leaves. And this is all that an automatic counter does; the series of boxes going down a conveyor belt, say, are put into one-to-one correspondence with a series of interruptions of a photoelectric relay and thereafter with the series of one-notch advances in a toothed wheel. What makes it counting is the recognition of one of these series *as* a counting series. That requires consciousness. The same consideration holds for adding and any other mathematical operation that the computer may perform. Computers can't compute.

Someone may protest that to say "The machine adds" must be to speak literally. For here is this column of figures. They have been added up. I didn't do it, and no other person did. But something did. Therefore, the machine did.—That is too simple a way of looking at it. You accept the last figure on the tape as the sum of the figures above it because you know or believe that the machine has been designed specifically to generate a series in correspondence with the counting series, composed of portions each as long as the separate figures in the column. The machine fits mechanical elements together in succession; you trust it to do this in such a way that the printout refers to the same figure that you would have reached had you yourself consciously gone through the fitting process item by item. So, in a sense, the answer to the supposedly rhetorical question, "Who or what added them, if not the machine?" is, nobody and nothing added them; they weren't added at all; but the machine has produced a figure which we know (or believe) corresponds to the figure that would have been reached had they been added; and that is all we are interested in.

This point may be more readily appreciated in terms of the abacus. No one would want to say that the abacus literally does addition; it is a tool to aid in addition. But an adding machine is only a mechanized abacus, with the advantage of printing out the "sum" in Arabic numerals. But the figures have to be interpreted *as* the sum, however easy a job of interpretation this may be. — But the skilled abacus operator does not think about every move he makes or even any of them; he just does it. — Quite so. I am not saying that in order for a process to be one of addition in the literal sense, it has to be consciously attended to at every step. I am claiming only that whatever the mechanics of the process may be, the intermediate or final configurations (bead positions, gear positions, marks on paper) have to be recognized *as* steps in addition; and this recognition, like any recognition,[13] requires consciousness. After the final catastrophe, the computers in the basement of the Chase National Bank may go on for a while, wheels turning, tapes being magnetized and demagnetized, marks being made on paper. But calculations will have ceased with the last gasp of the last clerk.

This is not to deny that "the machine adds" is, though a metaphor, a thoroughly dead one. If dead metaphors are to be considered as incorporated into the literal significance of words, so be it. The machine adds. But when we are considering the possible performance of insentient things, we will have to resurrect the metaphor or, if you prefer, invent a distinction of two kinds of adding, one of which presupposes consciousness while the other does not. And we shall then have to go on to point out that the latter kind of adding is only ancillary to the former and is meaningless without it. So we shall have arrived at the same conclusion.

The argument can be generalized and applied to most if not all the verbs starred on the list.[14] Granting that insentient devices cannot think—to use a summary word for intellectual activities—might they not, never-

13. Doesn't the radar recognize the approach of missiles and warn the countryside, just like Paul Revere? —No. —Then are you denying, a priori, the possibility of mechanical pattern recognition? —No. We must distinguish between recognition of a pattern and recognition of the meaning of a pattern. See the next chapter.

14. Some of them deserve special attention.

"Err," "mistake." Even though errors and mistakes can be and usually are unconscious, it follows from the argument just presented that an insentient being could not make an error or mistake, for the possibility of error can exist only where there is a possibility of getting it right; and the criterion of getting something right is always one of significance. There is no error in the *marks* $2 + 2 = 5$, nor in the mere event consisting of a more or less vertically descending sphere falling unimpeded to the ground in the vicinity of a man dressed in knickers.

theless, simulate such activities to such an extent as to accomplish all the aims of thinking? Silk purses cannot be made from sows' ears, but nylon bags can. What about the sci fi fantasies of "people" accepted on intimate terms into the best society who when involved in motor accidents spill out not blood and guts but springs and transistors? The gloomier stories have them taking over and disposing of all the real human beings, after which everything appears to go just as before, but there is no more consciousness in the world.[15] Are there any grounds for classifying these yarns as impossible in the same way that time travel is impossible?

Setting limits in advance to human ingenuity is a game with few winners. However, it seems that we cannot burke this issue, for if it is possible for a nonsentient thing to display all the outward signs of sentience, then the identity thesis, at least in the form in which I am advocating it, can hardly be sustained.

The question may admit of more fruitful discussion if narrowed down and made more precise. First, we

---

We say that the adding machine with a slipped cog makes errors; that is a *façon de parler* of the same sort as saying that it calculates. The norm from which it deviates is that of the sentient operator, not of the machine, which satisfies the eternal laws of mechanics just as impeccably in a breakdown as in "normal operation."

"Seek," "find." If you can't look for something without knowing what it is you are looking for, nor find it without recognizing it as your goal, then these words ought to be starred. And what are we to say of the missile with a heat-sensitive element steering it by means of servomechanisms toward a jet engine: Is it seeking the plane in the same sense in which a prospector seeks gold or only exhibiting superficially analogous behavior? I incline to the latter view but have left the verbs unstarred out of reverence for cybernetics.

"Translate." The virtual abandonment of work toward developing a computer program for translating between natural languages, after untold millions had been spent, confirms the point that translating has to do with meanings and is therefore not a suitable task for unconscious mechanisms.

15. This is the epiphenomenalist nightmare.

should peel off nonessential features such as visual appearance. There is no reason why we should require an intelligence simulator to be presented in a human shape. It need not even have a spatially unified body but might consist of subunits in various places, linked by radio transmitters and receivers. Nor need it literally speak; it would be enough if it could print out its remarks on a typewriter.[16] Nor should we demand that the machine duplicate human motions, even though some of these (those involved in brain surgery, for an extreme example) are just as significant manifestations of intelligence as any talk whatsoever. The machine is exempted because, not having human form, it could hardly move the same way.

Let us specify more explicitly and exactly than hitherto the kind of machine we are to consider, for, as we have seen, it is hard to draw any but an arbitrary line between machines and other kinds of things. Here we can turn to the standard literature. The device around which all discussion of artificial intelligence centers is what A. M. Turing calls the "discrete state machine"[17] and other writers call the Turing machine. The important notion of discrete states perhaps needs more explanation than it sometimes gets.

Every machine has working parts which change back and forth between two or more conditions in the course of the machine's operation. A part of a machine may be an electrical contact, which is either closed (state 1) or

16. In some natural language such as English. It would be all right for the printout to be in a machine language if that language could be translated into a natural language in accordance with an algorithm, that is, with no necessity for interpretation of machine symbols, only substitutions of equivalences. Such work is a typical computer task.

17. Turing's classic paper "Computing Machines and Intelligence" appeared first in *Mind*, 1950, p. 433–460.

open (state 2); or a dialing mechanism, as on a telephone, which can be in any one of eleven states. Or it
may be a pointer able to move continuously across a dial
face and thus be in an infinite number of states, which,
however, are not discrete; hence an instrument like this
cannot be an *operating* part of a discrete state machine.

A simple lamp is a discrete state machine with only
one operating part and two states, ON and OFF. A
3-way lamp, which contains two separate filaments, is
at any given time in one of four states. The number of
states in which an electric fan may be is infinite since the
blade revolves continuously.[18]

If all the operating elements of a machine are discrete
state elements, that is, they "move by sudden jumps or
clicks from one quite definite state to another,"[19] it is
evident that the machine as a whole moves in this jerky
way from one state to another. So it is possible to give a
complete description of the state of the machine at any
moment simply by listing the states of all the parts at
that moment. This is a bit of an idealization, like the
frictionless pulley. Some kinds of lamps go on and off
rather slowly, some very quickly, but even a neon lamp
takes some finite time to go from OFF to ON. There are
no absolutely discontinuous processes. But some, mostly electrical, come close enough to the ideal so that this
fact can be ignored. To speak more accurately, it is not
the speed, absolute or relative, of transition that deter-
mines whether a machine component is discrete state or
not but just whether the state of the machine at every
moment *that we are interested in* can be unambiguously
specified. This condition can be met, notwithstanding

18. But you could choose to regard its being ON as one state, OFF
as the other.
19. Turing, *op. cit.*, p. 439.

transition periods of great duration, if only the rest of the machine will, so to speak, "hold its breath" while the slow component is passing from one state to another. However, if this condition is not met, a machine will not be discrete state even if none of its components is capable of continuous variation. Consider a flock of sheep crowding through a gate several times wider than an individual sheep. We cannot count them, not because they move too fast, but because we cannot distinguish successive states of the flock which differ just in that one more sheep has passed through the gate. The states of this configuration are not discrete.[20] If we want to count sheep, we arrange for them to run single file through a chute. This can be regarded as a discrete state arrangement. However, if we run a flock of sheep simultaneously through parallel chutes, even though each individual chute is a discrete state, the summation of them is not.[21]

We can now turn to the most fascinating discrete state machine, the digital computer. This is a device for performing additions and subtractions[22] on any numbers within the capacity of the machine to handle, in any series of discrete steps desired, the specification of this series of steps being known as the program; and for

20. This is a magnified model of fluid flow and shows why "wet" apparatuses are not discrete state machines—an important point because the brain is "wet." Turing (*op.cit.* p. 451) says: "The nervous system is certainly not a discrete-state machine."

21. We could perhaps decide to count not "sheep through the gate" but "tips of sheep's tails through the gate," and then we would have a discrete state situation after all. This out, however, is not generally available to turn every configuration into one of discrete states, for we cannot, in general, decide arbitrarily what are to be regarded as the operative parts of a machine.

22. I here lapse into the jargon of computerology, without taking back what I said about the nonliteralness of attributing to machines the ability to compute.

displaying the result in a perceptible form. A mechanical desk calculator is a simple exemplar of this type of machine. The multiplication of 4 by 3 can be represented by the schema of Figure 2. Instruction I is put into the

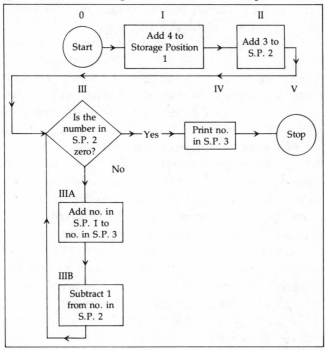

FIGURE 2.

machine by the operator's depressing the button marked "4." By a mechanical linkage, a notched wheel, which we can call "Storage Position 1," advances by four notches. Similarly for Instruction II. Pressing the "×" bar sets into operation the rest of the program, which is built into the machine. A bar moves into the notch in the S.P. 2 wheel; this is in effect answering "No" to the question in III, for if that wheel remained set at zero, the rod would have encountered the wheel itself, not a notch. So the subroutine IIIAB goes into operation. The rod is connected to a mechanism that advances the wheel S.P. 3 the same number of notches that S.P. 1 has already been advanced. S.P. 2 is moved backwards one notch. The rod engaged in S.P. 2 is withdrawn, moved back to its original position, and again probes the wheel S.P. 2. Since it encounters a notch once more, the wheel S.P. 3 advances an additional four notches so that it has gone eight notches from zero. Again the process is repeated, advancing the wheel S.P. 3 four more notches. Since there are only ten notches on its circumference, when it passes the tenth, a linkage causes another wheel (the "tens" wheel) to advance one notch. At this stage the wheel S.P. 3 has been moved back all three notches so that at the next probe the rod, encountering the wheel and not a notch, is unable to move further. The printing mechanism is thereupon activated so that the numerals "2" and "1" (reading from right to left) that have been brought into position by the rotations of the two S.P. 3 wheels, strike through an inked ribbon onto a roll of paper.

The same process could be carried out by three clerks at three desks handing pieces of paper back and forth, or by one person with three sheets of paper, or by a machine which stores figures in the form of magnetized patches on recording tape. These arrangements, and many others, however different they may be in their physical constructions, can all do the same thing: not "the same sort of thing," but exactly the same thing, namely, put numbers into and take numbers out of storage positions and transfer them from one storage position to another, according to the instructions they receive, which may, as we have seen, be of a conditional form, that is, do so and so if such a number is in such a position, otherwise do such and such.

The reason some digital computers cost $19.95 while others cost millions is that they differ in speed and in storage capacity. Cogs in a desk calculator may appear to move very quickly, but their speed is as nothing compared with that obtainable by electrical circuits, which have no macroscopic moving parts, hence negligible inertia. And while the one hundred ten-toothed cogs of a desk calculator can store only about ten bits of information, the larger computers have capacities in the millions.

One refinement that could be built into any calculator, but is in fact found only in large computers, is a randomizing element. The instruction at some point of a program may be: "Do this or that or t'other depending on whether the number in storage position so and so is 1 or 2 or 3." And the number in this position may be made to depend on some random sequence, such as the digits in the decimal expansion of π, or the series of next-to-last digits in telephone numbers as listed in the directory.[23] The principal usefulness of this accessory is in enabling samples to be taken where the total mass of data available for a certain problem is too large to be processed item by item.

These examples, and the history of the development of the computer from calculating machines as well as the fact that the main use of all computers is still as calculators, may suggest that the digital computer is only a Rube Goldberg device for doing sums faster than you or I would care to. But the computer is far more than that because the numbers that it manipulates can represent

23. A random sequence is one in which, of all the elements that can occupy a given place, none is more probable than any other.

anything we please. Thus, we can assign numerical
values to the letters of the alphabet, encode an English-
Italian dictionary, and store it in the computer. Then
when we want to know the Italian for "cat" we put this
word on the input tape with the instruction to print out
the next item after it in storage, whereupon the printer
produces "gatto."[24] Computers can play tic-tac-toe and
never lose because all the possible positions in the game
can be encoded, and the best possible next move for
each position is calculable. Computers also make a fair
showing at playing checkers and chess; that is, they can
be programmed to recognize which moves are legal and
on the basis of certain formulable rules of strategy can
assign probabilities of effectiveness to these and choose
the one with the highest value. The most interesting fact
here is that the computer can "learn"—if a move leads
to disaster, it can reduce the value to be assigned to that
move, hence do better on the next round. It can also,
like the human player, examine the possibilities for
some moves in advance and choose the one that is most
likely to lead to a winning or improving combination.
The number of possible ways that a chess game can go
in only three or four moves is so enormous that even the
largest computer cannot go through them enumera-
tively; but it can sample perhaps some hundreds.[25]

   These are only a few of the applications to which a

   24. Unfortunately it is not quite so simple a matter to translate "The
cat is on the mat," as everyone who has struggled with a foreign
language knows; for there are three Italian words for "mat," depending on whether it is made of straw, or is something to put under a
vase, or under a plate; and "on" can be either "su" or "sopra." Additional instructions could be given to the machine to enable it to cope
with these complications; but that is the bare beginning of the troubles
with machine translation.
   25. It is strange that with this potency no computer has yet played a
master game of chess. But wait and see, say the computerophiles.

digital computer can be put. The claim is made that the digital computer is a *universal machine*, that is, any digital computer can do anything that any discrete state machine whatsoever can do, considerations of speed and storage capacity aside.

This claim, interesting as it is in itself, takes on even more glamour when there is added the proposition that a digital computer can duplicate anything that an analogue (non–discrete state) computer can do; for the implication is that if the brain is in *any* sense a computer, then it can be mimicked by a digital computer. The impingement of this thesis on our conception of the human mind is striking. Three centuries ago Hobbes wrote:

> By RATIOCINATION, I mean *computation*. Now to compute, is either to collect the sum of many things that are added together, or to know what remains when one thing is taken out of another. *Ratiocination*, therefore, is the same with *addition* and *subtraction*.[26]

Moreover the brain is largely composed of electrical circuits that are either ON or OFF. This fact, together with the conception of ratiocination as calculation which is taken for granted in much modern philosophy, has seemed to render very plausible the view that the brain *is* a computer, hence that in studying the theory of computer circuitry and functioning and devising more sophisticated computer programs, we are studying the model of the mind and making it more accurate; in short, that psychology is becoming a genuinely exact science by this route. To enthusiasts of this persuasion, questions about sentience tend to drop out of discussion as unimportant or uninteresting or "to be left to the philosophers" or altogether meaningless.

26. *Concerning Body*, I, 1.

Thus Turing, dismissing the question whether machines can think as "too meaningless to deserve discussion," proposed to substitute "Can a machine be built which will successfully play the imitation game?" This pastime requires three players: two men and a machine. One of the men, and the machine, are labeled "A" and "B" and concealed from the other player, the interrogator, who writes questions on a typewriter, addresses them to A or B as he pleases, and receives their typewritten replies. The object of the game for the interrogator is to determine which of the two, A or B, is the machine. Both man and machine do their utmost to convince the player that they are human. Thus, for example, the machine, if asked to multiply two long numbers, will wait a suitable time before replying and might even "make a mistake" in the answer. Turing, writing in 1950, answered his own question with admirable exactness: that in sixty years' time (that is, by the year 2010) a machine could be built, with a storage capacity of about 10^9, that would win this game at least 30 percent of the time when the questioning period was five minutes long.

At the present writing, no machine can even begin to enter this competition. But Turing still has thirty-five years of grace.

Most of what can be said about this and similar sanguine forecasts of the the computerophiles has been said by Hubert Dreyfus in his book *What Computers Can't Do: A Critique of Artificial Reason* (1972). He summarizes the accomplishments that have been made to date in such endeavors as machine translation, chess playing, pattern recognition, and control of artificial limbs, showing in each field a pattern of initial rapid advance that is followed by a slacking off and a failure—over a comparatively long time—to follow through on the first successes. Machine translation projects, for example, started off bravely enough, and early in their history it seemed that only a little more refinement of programming techniques would produce readable translations of texts from the ordinary literature. But that little bit extra never was forthcoming despite the efforts of teams of ingenious and dedicated workers. For many years work has gone on in the endeavor to develop a reading machine, that is, a computer program that will enable the machine to take ordinary printing or writing as input when scanned by some suitable instrument such as a television camera. Here as well initial success with selected and specially devised patterns was followed by failure to generalize so as to cope with even a modestly representative selection of printing styles in common use, let alone handwriting.

Dreyfus explains these failures as necessitated by the nature of the digital computer.[27] This device was well named "the analytical engine" by its inventor Charles Babbage. Whatever it is to deal with must first be broken down into units, and it cannot ever consider a whole as anything but the totality of the units comprising it. This has the implication that whenever the computer is required to identify anything, it can do so only by going through an identification check list. Is it open in the lower part of the first quadrant and in the upper part of the fourth quadrant, this opening being continuous? Is it closed and continuous in the second and third quadrants? Is it free of intersecting curves in the continuous portion? If the answers to all these questions are affir-

27. He does not consider analogue computers to be necessarily equivalent to digital computers and does not predict what they may or may not accomplish.

mative, then it is probably a letter C, but only probably, for it might be a left parenthesis.[28] Does it have a cork-screw tail, a snout terminating in a sort of disc, and go oink? Then it is a pig, with probability 0.68. But this, as we have seen in Chapter II, is not the way *we* recognize pigs or letter C's. Nor is it at all plausible to suggest that we really do go through a rapid and unconscious list. We just recognize them. Later on, if required to do so, we may be able to discern and describe distinguishing marks;[29] but it by no means follows that those marks were the ones by which we recognized the beast or the letter in the first place, nor that we recognized them by marks at all. It seems that only people suffering from a certain form of aphasia have to check off distinguishing marks before they can decide whether a certain object is of a certain kind.[30]

In these examples we have been assuming that whether or not recognition proceeded via distinguishing class marks, at any rate there were class marks there. But this assumption is often unwarranted. To take Wittgenstein's famous example, there is no element that all games have in common that makes them

28. How is the computer going to decide whether this group of three is the word CAT or a parenthetical remark beginning with the word AT? Should it search ahead for a right parenthesis? For how many lines before giving up? And what if there is another C or possibly a left parenthesis between here and the right parenthesis that it finds five lines down? This is a sample elementary pattern recognition problem.

29. On the other hand, we may not. Consider the chicken sexers, who are quite unable to explain how they tell whether a chick is going to be a rooster or a hen, but they do it all the same. And amateurs of music can distinguish, say, Chopin from Schumann with perfect confidence and accuracy, though only musicologists would be able to specify distinguishing marks.

30. Hubert Dreyfus, *What Computers Can't Do: A Critique of Artificial Reason* (New York: Harper & Row, 1972), p. 35, citing Merleau-Ponty, *Phenomenology of Perception*, p. 128ff.

games. The only way to get the analytical engine to recognize a game, then, is to put into its information store a complete list of games: a game is acey-deucy or authors or . . . or Z-cars. If required, then, to decide whether something is the name of a game, the machine can run through the list and see whether the name matches anything on it—a somewhat unsatisfactory proceeding, apart from the watchfulness and expense required to keep the list current. Of more theoretical significance is the obvious fact that a procedure of this kind could not be called a recognition in any nonarbitrary sense of the word.

What, then, are the prospects for the great imitation tournament of 2010?

To be a starter, a machine must be able to accept programming in English,[31] for it would be cheating if the machine had a human programmer in attendance, ready to translate the interrogator's questions into Fortran or some such code. This means that the computer will have to "learn" the English language in the sense that it must be able to make all, or nearly all, the linguistic discriminations that a human speaker would make and use the language with human effectiveness. This amounts to saying that the computerophiles have thirty-five years left in which to discover the Linguistician's Stone—the function that will generate an infinite set of meaningful, and only meaningful, sentences in English. That is not all. The function has to be one that can be put into a computer, that is, it must be "analytic." This is to say, instructions must be prepared which will enable the machine to discriminate between combinations of English words that express a meaning

31. Or some natural language; but it might as well be English.

and those that do not. It is not enough that the machine should be able to distinguish between sentences in approved syntax from those that are not, for "Ain't none" can be a meaningful sentence whereas "Colorless green ideas sleep furiously" can't (or hardly can).[32] A syntactic criterion will not do, the machine requires a semantic one.

It has been suggested that an analytic engine might acquire language the way a child does—by trial and error, experience correcting the efforts of an innate aptitude. Let us not worry what the wiring in of the "aptitude" would consist in; let us assume that problem solved. We let the machine hear proper speech, and we let it produce expressions of its own, simply at random to begin with. We encourage it when it says something right and discourage the wrong expressions; how one rewards or punishes a computer is not clear, but presumably some kind of reinforcement of correct responses is a possibility—an adjustment of biases in a randomizing element, or something of that sort perhaps.

Unfortunately for the proposal, this is not and cannot be the way a child learns language, as Noam Chomsky has shown.[33] The reason is that such a procedure could at best (that is, ignoring contextual requirements) result only in correct repetitions of the paradigms. It cannot generate new, correct utterances—in Wittgensteinian

32. If, as I believe, it is not sentences as such but utterances in concrete contexts that are meaningful, the computer will be in even worse trouble. Supposing it somehow to have acquired a criterion of OKness for sentences, it would yet have to decide, on quite different grounds, whether the OK sentence managed to convey a meaning—and if so what—in each particular context of utterance.

33. For example in *Aspects of the Theory of Syntax* (Cambridge: M.I.T. Press, 1965) Chap. 1, Sec. 8.

terms, it cannot of itself insure that the trainee "knows how to go on." In short, the most that can be expected of the experimentation-and-reinforcement method of language learning, in child or machine, is the internalization of a phrase book. And unlike the traveler in a foreign land, who at any rate knows the meanings (in his own language) of the phrases he reads out of the book, the machine's ability to produce the phrases would have nothing to do with their appropriateness to context. This proposal could produce only a computerized parrot.

But assuming that somehow the engine can be got to produce a reasonable facsimile of English speech; still it needs to have something to talk about. At first glance, this might not seem to present a problem, as we all know that computers can have enormous storage and may be caused to print out the whole or any part of the *Encyclopaedia Britannica*. So if we ask it to give us a lecture on just about anything, it will oblige. There is likely to be some embarrassment, however, when we ask it for opinions. How is it going to answer the question "What do you think of Picasso?" Well, if it has the *Encyclopaedia* in its storage, it can look up Picasso, where it will find a magisterial judgment that it can print out. But its wily programmer will have instructed it not to plagiarize so blatantly but to paraphrase.[34] Even this, though, will hardly be enough to win a round in the imitation game. So the machine should rearrange the *Encyclopaedia*'s sentences, take some out, add some others, and disagree with some. How does one program a machine to disagree? Presumably, by negating the

---

34. To assume that it can do this is equivalent to assuming a solution to the mechanical translation problem.

sentence disagreed with. But here we begin to get into serious trouble. It will hardly do to negate something like "Picasso is one of the most important artists of the twentieth century." Disagreement to be at all plausible must concentrate on points such as "A basic continuity of artistic feeling can be discerned in the artist's whole production, despite the great dissimilarities between his many periods." The negation of that sentence, by text-book rules, would come out, "Either no basic continuity can be discerned (etc.), or he did not have many periods, or they were not greatly dissimilar"—a some-what strange utterance to make. And how would it fit into the rest of the revised computer judgment? In any piece of reasonably connected writing, you can't negate a certain number of the sentences arbitrarily and have the revision remain coherent. Nor can you negate all of them without falling into manifest absurdity.

A computer enthusiast might impatiently interject that we are only pointing out the difficulties in a machine's concocting an imitation opinion, and in doing so we are assuming without proof that the engine could not really have an opinion. Let us pay attention to this matter. One rejoinder, already made in the previous chapter, is that the having of opinions is something that within the conventions of our language cannot be attributed to nonsentient beings. We say that our opin-ions are expressions of the way we *feel* about things. However, let this be waived because it assumes in itself, perhaps wrongly, that a digital computer cannot be conscious. Let us return to the question how a computer might be constructed and programmed to have an opinion.

An opinion has to be about something, though not necessarily about facts; it sometimes happens that we have opinions about what we only mistakenly believe to be the facts. But let us ignore that complication—perhaps the machines are better than we and never get their facts wrong. Now facts,[35] no matter how combined and permutated, never of themselves generate any opinion, only more complex facts.[36] This is so even if we allow for combinations of facts to constitute so-called inductive generalizations. But the construction of a digital computer allows for nothing but the combination and permutation of stored data. So we are faced with this dilemma: Either the data stored in a computer represent facts only, and there can then occur in the computer printout no opinion; or some of the data are opinions. But if the data are opinions, they were introduced from the outside. Hence, though they may appear in the computer's printout, they are not the computer's opinions but something passed through it. No doubt many opinions held by human beings are of this sort too, possibly even all of those that concern Picasso, for example. But we can hardly be so cynical, or indeed illogical, as to suppose that every opinion is nothing but the effect of suggestion. The opinion that strawberries and cream are nice must often be autonomously generated.[37]

Let us remind ourselves of some fundamentals. A digital computer is an arrangement of physical parts, which can physically influence one another: They can

35. Including facts of the form: So and so's opinion concerning such and such is. . . .

36. This is what is right about the dogma "You can't derive 'ought' from 'is.' "

37. Turing, *op. cit.*, unimpressed by the contention that you can't build a machine to like strawberries and cream, snorts that it would be silly to try. But I suspect he would conceive of a strawberries-and-cream-enjoying machine as a device to emit the noise "Ahhhh" when strawberries and cream are the input.

push, pull, turn, and admit electrons into one another. In common with every physical system, it has the property that what comes out of it is a function of what goes into it and what happens inside it. The parts are put together in the way they are in order that a certain input will give rise to a certain output, the whole process being useful to human beings in compassing material ends.

The preceding is a beginning of a physical description of a computer. If it were made complete, we would have said, in a sense, all that there is to say, for from the complete physical description, together with a knowledge of the behavior of matter, we should be able to deduce what kinds of processes would occur in this thing under all conditions, including those of proper input. But the complete description would be a useless document; it would not merely be hard to deduce from it the *proper* working of the contraption, it would be impossible. For a mere physical description, limited to stating the compositions, shapes, and relationships of the parts, would include in itself no reference to uses or goals, would say nothing about what *ought* to be done with it, would tell us equally what would happen if it were connected up to a 250-volt main with a circuit fused for fifteen amperes, and what would happen if it were fed 220,000 volts, or attacked by rioting students with sledge hammers, or encased in plastic.

The repairman might have some use for a complete physical description, but even he would find more useful for most of his purposes, and indispensable for some, a different kind of description, one which states, or presupposes, the use to which the machine is "properly" to be put and explains how it works in teleological terms. Such a description is a wiring diagram, which names parts in terms of what they are *for*, such as transformer, resistor, input jack; in which the amount of attention given to details is determined by their relevance to the functioning. It will state (what the physical description will leave to be inferred) what the electrical value of a certain capacitor is but have nothing to say about its color or how many screws of what size are needed to hold it in place.[38]

We can draw a corresponding distinction between two descriptions of the input. One would describe it as cardboard cards with holes cut out of them or as plastic tape coated with iron oxide and variously magnetized. The other might characterize the input as "the census figures for Wilkes-Barre, Pa." Here we need to be careful. The latter description is more useful and more dangerous. There is not an exact parallel between this kind of description and a wiring diagram. A coil of copper wire is literally an inductance, but a punched IBM card is not literally a set of figures. There exists a relation between the figures and the pattern of holes, which is to say that there are rules for getting from the one to the other. These rules involve conventions. A certain pattern of holes in the card stands for "net increase of 7.6 percent in rate of illegitimate births during the decade 1961–1970," not on account of anything in nature but because someone has decided that it is going to stand for it. The same pattern of holes, chewed by silkworms in a mulberry leaf, would mean nothing, though it might produce the same effect if it fell into the machine.

So it is not correct to say that we store information in

38. From this we can see how superficial it would be to suppose that a wiring diagram "merely gives facts." Yes, but it gives a certain selection, out of the unbounded mass of the ideally complete physical description, and the selection and the nomenclature are determined by relevance to purposes—if one may be pardoned the expression, by value judgments.

the machine. We store something that stands for information. — A harmless figure of speech, it might be thought, no different from saying that books contain information—they contain marks on paper, but to anyone who can read, those marks *are* information. — Quite so. The point, however, is that while we are never tempted to suppose that books think or calculate, it is otherwise when we are dealing with something as dynamic and surprising as an IBM 360. So it is worthwhile to remind ourselves of the way things really stand.

With these distinctions and cautions in mind, let us return to the analytical engine trying to play the imitation game. What we call "asking the machine what it thinks of Picasso" consists of causing the first in a train of events. The machine has a certain program, that is to say, it is in a determinate condition, and the events consisting of the introduction of the input will set off others in a determinate sequence. The machine is constructed in such a way that every event in it can be represented by the addition or subtraction of a number to some other number already assigned to a certain storage unit; that is why all digital computers are equivalent.[39] Hence, to speak of numbers as being inputs and outputs of computers and stored in them is nonliteral only in the way that numerals are not the same as numbers; and since it will simplify discussion without leading to confusion if we are on our guard, we shall speak in this manner henceforth. Very well then, the input "What do you think of Picasso?" is trans-

39. And it is why there can be no improvement in digital computers *in principle*; they can only get bigger and faster.

formed into a series of numbers, which can be considered as one long number, the digits of which cause the machine to go through various operations, resulting finally in another number being either printed as the output, or pretranslated into some English expression such as "Picasso was the seminal painter of the first half of the twentieth century." How was the machine able to do this? Because the word "Picasso" and its place in the sentence are translated into numbers which stand for instructions to search[40] the storage for data including the word "Picasso" and to fetch those coded as "Picasso, opinions about." If there is something there, the machine will produce it; if there isn't, the machine cannot make it up. Its storage may contain the complete *oeuvre* of Picasso,[41] but unless there has been put into the storage, in addition to this, some critical remark, the poor computer cannot even say "I like it" or "Phooey." If it does say "Seminal painter" or "I like it," having been programmed to do so, the next move of the interrogator will be (as if he were examining a human student) to ask "Why?" Perhaps the computer will then haul out a stored essay on Picasso. But if any particular statement in the essay is queried, it will be unable to come up with a plausible, or any, reply; this is bound to happen in a very few steps. Maybe it would happen, too, with human students. But the good ones, in this kind of situation, would be able to place Picasso in his proper relation to the whole field of twentieth-century painting, to point out not only the similarities between

40. Again, these anthropomorphic words are used with the caveats expressed above.
41. Scanned by a color TV camera and digitalized.

"Les Demoiselles d'Avignon" and African masks but their significance in the context of revolt against the European rationalist tradition and attendant *nostalgie de la boue*.

I do not mean to suggest that the computer would sweat and collapse only when bullied with the esoterica of art criticism. The same kind of contretemps would develop if after asking whether it liked strawberries and cream, the interrogator followed up with "And would they be still nicer with a dash of ketchup?" A computer can "answer" any "question" by producing some statement already stored in it or the logical consequences of such statements. But it is not a logical consequence of any statements about strawberries, cream, and ketchup separately that the combination would be horrid. Therefore, unless the programmer expressly put this one in—and how could he anticipate everything?—the poor computer would be bound to give itself away.

Plato[42] deprecated the use of books on the ground that unlike teachers they can never answer questions but can only repeat what they have already said, over and over. Computers can do better than that because they can also state the logical consequences of what they have already said or are able to say. And no doubt those consequences can often be surprising, as Turing[43] testifies. All the same, this improvement over the book does not by any means extend to the possibility of the analytical engine's answering a huge range of relevant questions. The imitation game is and must remain, far

42. *Phaedrus* 275.
43. *Op. Cit.*

past A.D. 2010, beyond the scope of a digital computer.[44]

To say this is not to cock a snook at computers but at
those who would enter them in a competition for which
they are by no means fit.

Nor is it to say that the digital computer cannot model
some brain functions. That is for brain physiologists to
decide. But it is not to be supposed that the computer
can be the model of the brain; in particular, it gives us
no grip at all on the notion of sentience.

In the next chapter we shall try to say something
positive about this central conception.

44. Professor L. A. Zadeh in a public lecture has proposed a most
elegant strategy for winning the imitation game. The interrogator
begins by saying, "I shall talk for three minutes; when I stop, kindly
summarize what I have said." Any human player can carry out this
instruction with some success; no computer can, for to do so requires a
genuine grasp of significance, which is just what the digital computer
cannot have; see next chapter. (My concession, p. 105 ff., that a computer
could be made to understand a natural language and paraphrase sentences in it, was only for the sake of argument.)

V

SENTIENCE

In the preceding chapter we saw that we can do some things that a discrete state machine cannot closely imitate. This fact suggests[1] what should hardly surprise us, that sentience should be in some important and intimate way related to these abilities. In this chapter we shall try to find out what that relation might be.

In the first section we shall review some of the points made previously and begin to fit sentience into the scheme of things. We shall continue in the rest of the chapter with the nature and functions of sentience: the basic use of sentience as regulator of the motions of a mobile creature; finally, the sizing up function as the essence of sentience.

1. A CATEGORY FOR SENTIENCE

In Chapter II we argued that if sensations are brain processes at all, then they necessarily are and that the

---

1. Only suggests, because it was not shown that no nonsentient machine would be capable of these feats, in particular, the analogue computer. However, if analogue computers are equivalent to digital computers, they must be equally circumscribed in their potentialities. And no doubt the equivalence holds for all analogue computers which differ from digital computers only in passing continuously from one state to another. It may not hold for analogues involving holistic effects and what we might call "real" integration as opposed to arithmetical approximation, such as is exhibited in soap films used to solve certain problems in minimization. See Hubert Dreyfus, *What Computers*

statement "Sensations are necessarily brain processes"
is neither logically incoherent nor incompatible with
what we know by our experience of our own sentience.
Furthermore, if the claims of that chapter are acceptable,
there is no other, weaker version of the identity theory
of mind and body that is tenable. To say this much,
however, is not even to claim, much less to prove, that
the identity theory *is* correct. There remain as its rivals all
the multifarious dualisms from Pythagoras to Rhine.

What kind of rivalry is in question has often been misunderstood, on account of the history of the soul-body
distinction. At a primitive stage the soul may be thought
of as an internal double, or detachable organ, or fluid, a
material thing in its own right, which escapes from the
body at death. These theories are straightforwardly
"empirical," the evidence for them being circumstantial
and often convincing. However, when they reach the
degree of refinement where the soul is said to be a
thing, an individual unity, associated with the body but
not itself material at all, the issue has ceased to be an
empirical one. A new category unknown to previous
thought, that of *immaterial substantial particular things*,
has been proposed to include souls and perhaps other
entities such as gods, numbers, and Beauty Itself. The
question now is not whether we can discover these entities with our senses aided perhaps by instruments, but
rather whether we can talk intelligibly of them, whether
they can take some useful place in discourse that is
coherent and also compatible with obvious facts.[2] Thus,
to Lucretius it was a factual issue whether the soul could

Can't Do: A Critique of Artificial Reason (New York: Harper & Row,
1972), p. 233f.

   2. If you do not shrink from denying obvious facts, then, of course,
anything goes. If you say that all is illusion, or sense data, you can go
on to create whatever world you choose.

be identified with a certain kind of atoms, and his arguments, mostly citing concomitant variations in the soul
with changing bodily conditions, were appropriate for
the thesis he was upholding with due consideration to
the context of the dispute. However, in modern
philosophy since Descartes, the parties have not dis-
puted seriously over the brain and its manner of functioning. The controversy has been between those who
claim that the soul or mind can be understood simply as
this functioning matter, and those, the dualists, who
deny this and conceive of brain functions as correlates of
mental activities, which in themselves are occurrences
apart from the brain—either events in a nonmaterial
substance or perhaps self-standing basic existents.
Thus, the dispute, carried on in a day when brain physiology is a tolerably well advanced branch of knowledge,
is still philosophical, being about the nature of a conceptual scheme.

At one point in history the sentence "Sensations are
brain processes" could have been used to announce an
empirical or scientific discovery. It would have
contrasted with "Sensations are heart processes," "Sensations are the boiling of the blood," and the like. That
is not the situation now. The contrast is with "Sensations are ineffectual nonmaterial accompaniments to
brain processes," "Sensations are the data of which
brains are logical constructions," and similar pieces of
metaphysics which envisage a world of which the statements of physics and physiology are either not in principle a complete description (dualism) or, though allow-
able as *façons de parler*, do not literally describe anything.
An analogy, partly historical and partly fanciful, may
help to show why "Sensations are brain processes" can
no longer state an empirical scientific discovery.

The ancients once believed fire to be one of the elements—a substance, a thing, one of the great world divisions on a par with earth, air, and water. When they discovered that fire is not a thing alongside other things but a process, a combining, in fact, of "earth" and air to produce water and more earth, they could have announced the breakthrough as "Fire is earth-air-water processes." Further investigation in this field was devoted to finding out the particulars of the process. But if interests or emotions had been sufficiently bound up with the old view, its defenders could have claimed that the discovery had been only of the concomitants of fire, not of its real nature. Granted that in this world, they could have said, we don't find fire except where earth and air are combining to produce water, nevertheless it is evident that the nature of fire is repugnant to the cold. It is in itself pure heat. We certainly can conceive of it as existing apart from material dross and indeed, when thus apart and in itself, it must be even hotter.

This sort of controversy, between thermal materialists and idealists, could not be resolved by making more "empirical discoveries," at least not in any straightforward, crucial-experiment way. If it had occurred, though, it might have come to an end when a comprehensive kinetic theory was worked out. When everything can be accounted for satisfactorily in terms of the unifying notion of kinetic energy, it becomes psychologically difficult to hold out for a separable and immaterial heat, even if the logical situation does not veto this course.

The moral is that the way to put an end to fruitless controversy over mind and body is to produce a comprehensive theory of mind. But that is not so easy. Let us see, however, how far we can go in that direction.

Dualism nowadays exists, as far as I can tell, mainly in the forms of parallelism and epiphenomenalism, which are hard to keep distinct and perhaps really aren't. It hardly matters, for they share the root error of taking for granted that sensations are things, individual entities on their own, different from bodies, therefore forbidden by physics to interfere in the motion of bodies. So they must either constitute a series of their own, parallel to but having no effect upon the series of bodily motions— which is an incredible position; or else they have to be conceived as a sort of nonenergetic by-product or fluorescence of bodily motions, a thought which implies that if somehow they were annihilated, everything would go on just as before. People would no longer feel anger, or anything else, but when their eardrums were activated by those vibrations whose symbolic representations were once considered unprintable, adrenalin would still be introduced into the bloodstream. This is also an incredible position. The only alternative is monism, and to repeat it once more, the only coherent monism is one of strict and necessary identity.

Those who wish longer arguments in favor of a monistic position will have no trouble in finding them. But I prefer to direct effort toward explaining what does after all need explaining, namely, what is the use of sentience? If the world is just atoms and the void and nothing does or can happen but the banging of atoms against one another in complicated but regular ways, what is the point of there being creatures who are conscious of it? Granted that you and I are not analytical engines, still we must be mechanisms of some sort. A mechanism is just a way of getting from one condition to another, by way of intermediate lawful processes. Given the input, the output should be in principle calculable; or if it isn't, but is more or less random, that only accentuates the difficulty of imagining any reason why this process should be in part conscious of itself. But "nature does nothing in vain"; or if that sounds too medieval, then "evolution establishes no mutations without advantages." If one continues to follow this line of thought, one may be tempted to conclude that we aren't mechanisms after all. But in the broad sense of "mechanism" here in question, to say that we are not would be so far to give up the hard-won vision of a world whose principles of operation are understandable and to sink back into a morass of mystification. Let us try to avoid that.

A prophylactic point should first be made. The reifying of mental entities is facilitated by talk of "sensations." I have pointed out already, as Ryle and others have, that this word's ordinary employment is much extended in philosophy. The danger is that it serves as a convenient label for a "mental thing," gratuitously creating the impression that there are mental particulars, individuals, on a par with particular bodies. The word does this because in uses outside philosophy, it applies to features that are literally sensational—burning sensations, vertigos, goose bumps: standing out from the background consciousness with rather sharp beginnings and ends. A headache isn't called a sensation. An ordinary person leading a tranquil life may go for weeks at a time without having a single sensation, in this ordinary usage of the term. But philosophers say one begins to have them instantly on awaking, continuing all day, ceasing only when one falls asleep at night— or not even then, according to some authorities. It is rather as if philosophical oceanographers, noting that we have the word "wave" to signify an individually noticeable feature of water, first concluded erroneously that a wave is an individual bit of water and then went on to speak as if every body of water really was made up of waves. But in reality there are no mental particulars at all, short of individual minds, though of course there are discriminable features which we may arbitrarily bound and name.

If we have to have metaphors for the mind, then probably the hydraulic image of a stream of consciousness is best or least bad. And like another famous stream it just keeps rolling along; it doesn't *do* anything. This may not be obvious at once because sentience may be a cause or reason: "The pain caused me to scream in agony," "The thought that I might soon run out of money was the reason why I walked instead of taking a taxi." But the pain did not produce the scream—*I* did; nor did the thought of penury set me in motion—*I* decided to walk. Just because X is the cause of A, it does not follow that X must do something to bring A about. Static entities can be causes as can emotions, abstractions, and negativities, to name only a few. The ring of the Nibelungs caused the destruction of Valhalla, 'twas love that caused King Edward to lose the throne, barbarism and religion caused the fall of the Roman empire, and starvation causes death. But rings, love, barbarism, and starvation don't do anything. They are not agents. The notion of cause is altogether distinct from that of agency.[3]

This fact enables us to see what is right about epiphenomenalism, namely, that sentience is not executive, while rejecting what is wrong with it, the paradoxical

3. Agents are not always causes either. The child asked for the cause of Charles I's execution will not get marks for answering "the axe" or "the executioner."

contention that there cannot be mental causes. If we can understand how my anger can be the cause of what I do, without forcing me to do it, we may be on our way to resolving some of the puzzles about free will.

2. SENTIENCE AS REGULATOR

Imagine that we are called upon to design a device, the only specification for it being that it is to be put out into the world and must *last*. What shall we do?

The simplest thing would be to try to make something that would not wear out. We would use inert materials of great strength and hardness and be careful about the joints. We might thus succeed in creating something that would decay very slowly.

A more elegant solution, however, would be to produce something capable of repairing itself when it needed to and able to replace itself by something of the same kind when repair was no longer practicable. This would not simply sit and ward off blows; it would do something about maintaining itself and its kind. To be active, it should not be made of inert materials but, on the contrary, from reactive stuffs.

If it was to be put out into the world, into a random environment, it would have to be designed to use the materials at hand to repair and replicate itself and supply the energy needed for these processes. Therefore, it would have to be made out of abundant and accessible materials and to have some way of extracting the ones it needed.

Now we could make this device so that it would stay wherever we put it. Then it would have to make do with the materials within its reach. If they were insufficient, it would not last. Since it would be a matter of chance whether the necessary supplies were within reach, this might seem a bad form for our project to take. However, it would not be impossible to proceed in this manner if we did two things: made the device out of very common materials so that the probability of its finding the necessities within reach would be fairly high; and made enough prototypes and scattered them about to increase the probability that some of them would be in favorable environments.

The device would need an intake system capable of admitting what was needed while keeping out what was harmful. As it would not go after these materials but wait for them to come to it, the arrangements could be passive.

If the environment of the device was likely to contain some maintenance-and-repair substances in one direction and others in another, orientation mechanisms ought to be built in to assure that, for example, the sunlight receptors pointed up, the mineral intakes down. Some kinds of action involving motion would not be incompatible with the general specification that the device was to remain in one place; it might keep its sunlight receptors pointed toward the sun, for example. There might even be a mechanism for holding on to nutrients that would otherwise flee the scene before ingestion.

Chemical and mechanical devices to perform all these functions exist. And there is no need for plants (which we have been designing ex post facto) to have sentience or anything like it. This is so even of that extraordinary vegetable the Venus's flytrap, which can count up to two in the way a computer can "count": its hair trigger has to be stimulated twice before its petals will close. But they will close on anything that touches the hair in the right way. This is all right for the plant because in its usual environment, whatever stimulates this mechanism is likely, though not certain, to be a fly.

Now let us try to design a lasting device not required to stay in one place. Ability to move can be an advantage because if replacement parts and fuel are not available where it happens to be, it need not perish, as the plant must, but can move on to where there are better pickings. The advantage will not accrue, however, unless the device has some way of distinguishing one place *as better* in this respect than another. It may simply scurry about at random, but if it comes upon sustenance, it needs something to cause it to stop and ingest before moving on. This could be an automatic mechanism; the device's intake structure would continually test whatever it encounters for edibility, and whenever the result of the test is positive, a shutdown would operate on the motor while the material is taken in.

If the mobile device is to last, however, securing sustenance is not enough; it must also avoid dangers. Since it will not be built very solidly, it could be crushed by things falling on it, could fall into holes too deep to get out of, could be decomposed by excessive heat, or could get too close to the intake mechanism of some similar but larger device. We cannot hope for the success of the project, therefore, unless we can build in a danger-avoidance mechanism.

How are we to do that? We might begin by trying to list all the dangers likely to be encountered and provide our device with means for testing for their presence, with a positive result to initiate motion fromwards. This way we could build in analogues of reflexes and instincts. A thermocouple hooked up to the steering mechanism would reverse the course whenever a region of too great heat was approached; probes on the front could keep the device from running into obstacles or over precipices; and so on. How far? However unwieldy the device might become, it may seem that in theory we can carry this idea to any length. But already the machine is beginning to look like the White Knight.

Restricting the danger-avoiding mechanisms to those that operate by contact with the source of danger is inefficient, negating most of the advantages of mobility, for a thing that cannot get away from danger until right upon it will have to be very cautious in its movements and heavily protected. We need an early warning system.

There can be information at one place about the existence, character, and whereabouts of something at another place only in virtue of some effect of the latter capable of traveling over the intervening distance. Such effects may be in the form of chemical effluvia, sound waves, electromagnetic radiations (including reflections), gravity, and perhaps others. Detection mechanisms operating on the basis of these kinds of effects could be hooked up to our device to cause it to depart when something identified as threatening was in the offing. It is important to notice, however, that at this stage the interpretation of the effect as issuing from a threatening thing, indeed the whole notion of a threat, is something pertaining only to us, the builders of the device, and not to the device itself. For example, suppose the device is vulnerable to being devoured by tigers. We guard against this danger in the following way. We record the sound of a growling tiger and analyze it acoustically. Suppose we find it to consist of a vibration with a fundamental frequency of forty hertz with a first harmonic of 13 percent relative amplitude. We then design a filter for this combination of frequencies and attach it to our device so that when energy passes through it, the avoidance mechanism will start up. Then the device will run away from tigers—at any rate, when they growl. If they don't, something else will have to be tried.

Maybe we could do better than that, building in a single multipurpose sound analyzer that would discriminate between tigers, tanks, nightingales, the mating call of a similar device, and so on. Yet these would still be only so many inputs; they would not have of themselves any significance. We, the designers, would have to preset the device's responses, as flight from some, indifference to others, approach to still others. Presumably we would do this by building into the device a digital computer with a storage containing paradigms of these sounds, associated with appropriate response instructions. The input sounds would be matched up with them, and the device would move accordingly.

Perhaps we could in this way furnish our device with "ears" and a repertoire of appropriate responses sufficient to keep it out of the way of tigers, elephants, and motorcars with standard pitch klaxons. For many things do have characteristic sounds, and one instance of a tiger's growl or elephant's trumpeting is much like another. However, providing "eyes" for the device would be different. Here we would run once more into all the problems of pattern recognition. There is just no standard tiger shape, considering all the angles and distances from which a tiger may be seen and the postures it may be in. There is no template, nor set of them, that we can store in the computer with instructions that anything matching one of these is to be considered a tiger. Nor, having passed that hurdle somehow, could the device distinguish between tigers in the jungle, which are to be fled, from tigers in the zoo or circus or on the movie screen.

Not to labor these points further, since they follow from the remarks on machines in the previous chapter, we can say on the whole that an attempt to produce a viable animal on the principles available for constructing ordinary machines, including digital computers, is not in principle possible.

Nevertheless, there are organisms, assemblages of matter, which when put out into an environment at random, manage to find their food, evade their enemies, repair their lesions, and reproduce themselves. So although we do not know how to do it, by the Frankenstein Axiom it must in principle be possible to build such a device. Let us try to enumerate some of the respects in which it would differ from the inept and mechanical machine we have been imagining.

All animals share a single kind of immediate danger-avoidance system: pain. When we come up against something that is actually and grossly damaging us, it hurts,[4] and we have a strong tendency to withdraw. This is sentience. It works. However, in some kinds of dangerous situations, the withdrawal (from a hot object) or protective action (eyelid closure), being so fast that pain is not felt before the response is complete, cannot be on account of the pain. But if this can happen in some cases, why not in all? What is the point, we may ask, of the unpleasantness of pain if our safety can be assured by other means—in effect, by something like the automatic nonsentient mechanisms we were just now inventing for our movable device? There is no need for a fused electrical system to incorporate anything like pain; the fuse is enough. Whatever the necessities of sentience may be, we have here one that we might well dis-

4. There is one exception, which in the vulgar sense proves the rule: massive gamma ray irradiation.

pense with, or so it may seem.—I believe it is possible to reply to this kind of complaint against the economy of nature and will try to do so in due course.

Pain has at least three uses.

First, it informs us where and what kind of damage we have suffered; or at least it tells us where to look.

Second, it stimulates us to disengage from the damaging object or situation, though, as we have just noticed, we may sometimes do so before feeling the pain.

Third, it reminds us of unrepaired damage and keeps us from damaging ourselves further while the repairs are proceeding. For example, it prevents us from trying to walk on a broken leg.

Of these ends, the first and second might seem to be achievable by less unpleasant methods. Why could we not have a merely neutral feeling, like our kinaesthetic and orientational sensations, calling our attention to a crushed thumb? Besides, as a matter of fact, really massive damage is not immediately painful. Soldiers who have had limbs suddenly severed may go for some time without noticing the fact, we are told.

Part of an answer to these questions can be given if we ask ourselves whether it is conceivable that we should be afraid of a certain kind of situation, though knowing full well that if we found ourselves in that situation, we would experience only neutral sensations; for example, that we should be genuinely terrified at the notion of putting our hand into a buzz saw if we knew that the sensation of doing so, and its aftermath, would be distinctive but neutral in feeling tone, like sticking one's hand into a bucket full of ball bearings? I do not think this is conceivable. We might regard the situation as one to be avoided because we would *know* that sticking the hand into a buzz saw results in temporary or permanent disablement, which is disadvantageous and unpleasant. But a mere intellectual appreciation of a situation as one to be avoided is not of itself an incentive to avoid it, except possibly to perfectly rational beings, and then only because those beings are cognizant of the likelihood of future pain or because they fear death or disablement, whether or not it is painful.

This, I say, is *part* of an answer—that creatures like us are as a matter of fact motivated most effectively by fear of pain, in most cases for our own good. There might be other kinds of creatures with different psychologies, who would have only neutral sensations where we have pains, but who would fear those neutral sensations just as intensely as we fear real pains; or if we did not want to call that kind of attitude fear, still they would make it their first priority to avoid those neutral sensations[5] so that it would come to the same in the end. The rest of the answer can be given only when we find ourselves in a position to show what repercussions that kind of psychological difference would have on the rest of their mental economy.

However, when it comes to the third use of pain, that of reminding us of unrepaired damage and preventing us from damaging ourselves further, we can hardly suppose that a nonpainful substitute sensation or other mnemonic would do the job for any kind of creature. Pain, just by being what it is, tells us that it ought to be avoided. It is the Ponte Vecchio from Is to Ought. No neutral sensation could take on this function. If, when I try to walk on my broken leg, I experience some feeling on the order of tall grass brushing against me, I may be

5. Perhaps this is not so fantastic a supposition. People can be induced to have intense fear even of pleasant sensations, as history shows.

reminded that it is a danger signal and desist. But then again, I may not; I may forget or, though remembering, decide, for good reasons or bad, to go on walking nevertheless. But the actual pain will not let me forget, and its intensity, if it does not absolutely preclude my walking, insures that a decision to walk will be founded on compelling reasons, for instance, that unless I do, I shall surely be eaten by a tiger.

Whether or not some nicer arrangement could do what pain does, there is no question but that it does perform the functions mentioned, on the whole, efficiently if brutally. Compared to this big stick, the few carrots that nature provides have little significance. We should be grateful for them, but we need not discuss them separately. The ancient hedonists who defined pleasure as the absence of pain knew what they were doing. There is no neural network for pleasure as there is for pain, and indeed there are no sensations of pleasure, though some sensations are pleasurable.

Pain is basic to all animals, just because it is essential for getting around in the world. Without pain there is no possibility of action, and without action, no knowledge, not even of two and two are four. It is pain that establishes the existence of the "external world." There are no dream pains.[6]

But while there can be some prima facie doubts about whether pain is the best damage-warning and avoidance system, there can be none about the sentient system of distance receptors in comparison with the other

6. If there were, we would be afraid to go to sleep. For a dream pain would *be* a (real) pain. Of course there are dream anxieties, embarrassments, and the like. But dream fires never burn us, dream falls never break our legs, nor do they even seem to. Pinching ourselves to test whether we are awake is a perfectly valid procedure. If it hurts, then we're awake all right. It couldn't seem to hurt and not really hurt.

hardware that we tried to devise. We don't know how we do it, but we, and lowly creatures as well, easily solve the problems that baffle the designers of analytical engines. I say we don't know how we do it, but we do know something about how we don't do it. Consider item finding and pattern recognition by vision. There is one gross difference to note at the outset. Both we and a computer with a TV camera input "scan" the field, but in very different ways. Computer scanning, like all its operations, is sequential, that is, it proceeds a unit at a time. The camera mosaic is translated into voltage differences, which are fed into the computer one by one; the scanner moves across the mosaic, dot by dot in the top line, then back to the beginning of the second line and across, third line, fourth, and so on. The scanning can proceed in different orders, for example, radially as in a radar display, but always it has to move element by element. This is not the way *we* scan, as we all know from trying to search "systematically" for something in just this way. If we have lost a diamond ring on the lawn and cannot find it immediately, we finally stop beating about and go at it "methodically." We start at the point where it was most likely to have been dropped and follow a spiral pattern, riveting our attention to just one narrow band unfolding outward. Anyone who has tried this—or the alternative way of looking along a six-inch band at the edge of the lawn, then moving in to another narrow band, and so on—knows how hard it is to do and how much concentration it takes. For this is not our natural way of looking for things. Our way is to take in the whole at one glance and narrow down to a more careful scrutiny of some part of it suggested to us by various cues—the mechanical model being cinematography with a zoom lens.[7] This will not do for the ring on the lawn, but for most of our purposes it is fast and efficient.

Consider also the difference between the way we and the machine identify objects. As we have noted, there does not exist a routine whereby a machine can identify a tiger. If there were, however, its general character would be that of correlating the outline and other features of the tiger with a "template"—stored data on what characters indicate a tiger in the offing. That, again, is not the way we do it.[8] We simply see it as a tiger. There will be a certain sense of familiarity if we have seen tigers before, which will be heightened if we happen to know (that is, remember) that "tiger" is what it's called. But we do not compare anything with anything, nor remember (necessarily) any other particular occasions on which we either saw tigers or heard the word used. Recognition *cannot* be the result of comparison with a mental image because there are no mental images. But even if there were, the doctrine would have to be rejected. If I identify this approaching beast as a tiger by comparing it with my mental image of a tiger, I must first identify that image as the image of a tiger. And how am I going to do that? By comparison with the image of an image? But, you say, I know this image is the image of a tiger because it's the effect of my having

7. It is worthy of notice that zoom lens techniques are easy for us to follow and seem quite natural despite the fact that our ocular equipment includes no zoom device. But no movie director has yet had the audacity to scan his field line by line. In fact, we have difficulty following even slow pan shots. This is because when we move our eyes across a broad field, thinking that we are sweeping with our gaze, we are actually proceeding by a series of discrete fixes.

8. Though philosophers have not been lacking who maintained that we *must* compare our "impression" with a stored "idea."

seen a tiger. However, if we are going to stop the regress that way, we would do better to stop it at the first step, not the second: I recognize this beast as a tiger, my ability being the effect of my having previously seen a tiger.

Or consider art. The Japanese painter makes two or three squiggles with his ink brush, and it's unmistakably a tiger, not a house cat nor even a lioness. But any attempt to show topological similarity between the tiger and the drawing—supposing that we even knew how to begin to make the comparison—would produce little positive correlation. In caricatures an outrageous distortion of Nixon's features may be instantly recognizable, even though a detailed and fairly accurate portrait done by an incompetent artist might generate confusion with Andrey Gromyko.

However, our distance receptors were not given us for entertainment or even the gathering of disinterested knowledge but as aids in getting our dinners while avoiding being someone else's. It is in zoos that we see tigers merely as tigers; in the jungle we see them as dangers. And that is something of a feat. How remarkable it is will be appreciated if we recall again to what shifts we were driven in trying to conceive how to program the mobile device to detect dangers and how, even if it was equipped with a list, it would be at a loss to distinguish between tigers in zoos and tigers in jungles.[9]

Here are some important differences between animal and machine interpretation of data provided by distance receptors.

9. It could be instructed that tigers in zoos are not dangerous. But a tiger walking down the central esplanade of the zoo *is* dangerous. Is the computer to have that fact stored also?

1. Animal interpretation takes account of context; machine interpretation does not.

2. Animal interpretation goes from the whole down to the particular; machine interpretation goes from the elements to the whole, considered just as the aggregate of the elements.

3. Animal interpretation takes account of relevance; machine interpretation treats all data alike, unless instructed in advance and in detail which, in particular, to ignore or deemphasize.

4. Animal interpretation, generally speaking, is concerned with the data field as a field for possible action and considers it from that standpoint; machine interpretation, in general, does not.

5. Even the roughest and most preliminary animal interpretation is in terms of values—a special effort has to be made to "stick to the facts alone." A field that affords no values in its interpretation is by definition boring, and we cease our interpretative endeavors unless forced to continue them. Machine interpretation, on the other hand, discovers no values at all.

6. Human interpretation results in an assessment of the meaning of the situation interpreted. But meaning is a notion foreign to the machine vocabulary.

The items on this list overlap, and I make no claim that they exhaust the significant differences. My purpose here is not to present a complete array of the differences but to say enough to give some content to a notion I want now to introduce, that of *sizing up*. I would summarize the differences in the array above by saying that when an animal considers a data field, it characteristically *sizes it up*. But no machine does, nor can, if it lacks sentience.

3. SIZING UP

Let us consider jokes.

There is no use asking an artificial intelligence researcher, "Can you program your computer to get a joke?" He will only be irritated and dismiss you as a frivolous ignoramus. In a way, he would have a right to do so, for getting a joke—appreciating it, seeing the point, laughing—is not what artificial intelligence research is all about. Nevertheless, it will not be beside the point for us to make a more modest inquiry, asking not whether a computer could get a joke in the full sense but whether a discrete state machine with however large a storage could *detect* a joke (that is, from miscellaneous stretches of discourse fed into it, select all and only those that are funny), even allowing, say, a 50 percent margin for error.

The answer to this question must be negative. For all the machine could do would be to examine the remarks for the presence of certain traits that either constitute or are correlated with the quality of being funny. But there are no such traits. There are no particular forms of expression such that anything put into them must be funny. Nor, supposing *per impossibile* that the machine is able to grasp meanings in general, could it identify wit or humor from the meaning of what it encountered. Consider the remarks (1) "The best laid plans o' mice and men gang aft a-gley," and (2) "The best planned lays o' mice and men gang aft a-gley." How in the world is the poor machine to tell which is the platitude and which the takeoff? What template can it fit them against? It is worse off even than Bertrand Russell's German, who when reading the comic issue of *Mind* was able to syllogize: "Everything in this journal is a joke; this advertisement is in this journal; therefore, this advertisement is a joke."

The objection might be raised that we don't have really much grasp of wit and humor; if we did, some more or less mechanistic theory[10] might turn out to be correct, on the basis of which we might write a machine program for detecting or even inventing jokes. But this is a mistake. It isn't that we don't understand how the joke works—we do, well enough; the trouble is that we cannot use that understanding as we can when studying the workings of nature to generalize to a regularity that will aid us in prediction and control. Slight alterations in famous sayings can produce ludicrous effects; we all know that; but we can't use that knowledge to sit down and excogitate jokes ad lib. If you think you can, open up *Bartlett's* and begin. A computer could do it faster and might sometimes even come up with a joke; but it would be like Eddington's typewriting monkeys. And if the computer did produce a joke, it could not separate it from the nonjokes.

Part of the explanation lies in the fact that much wit and humor depends on contextual incongruity;[11] perhaps all comedy is sitcom. Generally speaking, the phrase "Bugger Lyme Regis" is not amusing, but it is different when we are told that these were the last

10. Hobbes, Bergson, and Freud have propounded theories, which despite their purported generality, account only for our amusement by moron jokes, people slipping on banana peels, and dirty jokes, respectively, and only dubiously for these.

11. This is the "derailment" theory of Max Eastman's *Enjoyment of Laughter*, which is more satisfactory than those of his eminent predecessors.

words of King George V. Our conception of what a
royal deathbed scene ought to be like is, as it were,
exploded by this intrusive detail. Yet not all unfulfilled
expectations or incongruous juxtapositions are
humorous; a flea and an elephant, merely juxtaposed,
are not funny, nor balls of different sizes.

Getting a joke is an instance of sizing up. For whatever else it may be, at least it is a process of perceiving a
complex field and noting that within the complex a certain item has an interesting relation to the whole or to
other items. Now "interesting" is an evaluative term,
not reducible to an alternation of terms specifying factual relations, such as "taller than" or "reversed" or
"especially-concerned-with-Scottishness," etc. How is
it that sentience is able to bring off this feat of dividing
things into the interesting and the boring? Because of
the primordial interest that all sentient creatures have in
finding something to eat. A machine could get jokes all
right, if only it could get its dinner—out in the world, on
its own.

A borderline example: tennis.

There is some plausibility in the notion that a tennis-
playing machine could be built. For it ought to be possible to design a mechanism that could calculate the tra-
jectory of an approaching ball in flight and move a
racket holder to intercept and whack the ball with the
right force and at the correct angle to send it to that part
of the court most awkward for the opponent. So
nothing but expense, more pressing projects, and anti-
mechanistic prejudice keep IBM from the Davis Cup.

What then is borderline about the example? This, perhaps: While the activity of tennis may be simulable by
computer, it could not be duplicated, no more than air-

planes duplicate the flight of birds.[12] The human player does not play in the way just described; he sizes up the situation and adjusts his strategy accordingly. The human player will lob a shot back to a side that ordinarily would be an easy return if he happens to notice that his opponent is momentarily off balance.

This illustrates the fact that the way of sizing up is not the only way in which certain things can conceivably or even practically get done, as the phototropism of certain plants likewise does. It should not be surprising after all that the animal soul can do some of the things that the nutritive soul also does, and vice versa; likewise the computative soul. There is no great gulf fixed between them; it's just that there is territory beyond the vague boundary of overlap.

Tennis is rather near the computer end of a spectrum of sports which for historical reasons has tic-tac-toe at its extreme. Bowling likewise is a completely computerizable or rather mechanizable game, though then it would lose all its point. We may assume that in all ballgames, robots could handle the ballistics. But baseball, football, and basketball introduce aspects of tactics and strategy where it seems unlikely that robots would be able to cope. As for what has been called the greatest game of all, one could perhaps take the Vietnam war as having shown, in a practical manner, the shortcomings of the robot general. If the bane of the military mind is its proneness to fight the last war, the defect is aggravated in the robot, which is limited to two procedures: following instructions which have worked in the past or "learning by experience" in the crude sense of elimina-

12. Someone has observed that airplanes come closer to duplicating the flight of beetles.

ting responses that have been tried and have failed—but which just might work if tried again in different circumstances.

Boxing and fencing, in which sizing up in a narrowly literal sense is all important, are at the other end of this spectrum, being the sports (one may conjecture) most resistant to computer simulation not only on account of the notorious difficulty of duplicating by computer program the actions of a human limb but also the skills required for recognizing feints for what they are. And these sports are close to the kinds of activities necessary for the survival of carnivorous animals.

More examples:

1.[13] Recognizing that in the sentence "The cow was chewing her cod," the word "cod" must be a misprint for "cud." I have been told by a computer enthusiast that a computer with enough storage, fed random information for nine or ten years, ought to be able to detect this kind of error. With all deference to expertise, I believe this prediction to rest on failure to grasp the point, which is that probably none of us, before the invention of this example, was ever taught, or heard anyone say, that cows don't chew cods; nor even that they don't eat fish; nevertheless, all of us would instantly recognize the error for what it is.

To give the computer a run for its money, let us suppose that it "learned" that cows are herbivorous and that whenever it "accepts a proposition" it rejects everything logically incompatible. Then, if it came upon this sentence, it would recognize it as *false*—but hardly as *absurd* and due to an error in one letter. Nor could it

13. Adapted from Gunderson, *op. cit.*

cope with a fantasy about ichthyophagous cows, in which "cud" might be a misprint for "cod."

2. Grading an essay, indeed criticism generally.

3. Historical explanation, which has little to do with causal laws in the sense of the physical sciences but everything to do with retrospectively sizing up the situations in which people found themselves so as to be able to appreciate why they acted as they did. And as success in getting a joke does not guarantee us the ability to think up new jokes, so success at understanding past events does not make the historian into a prophet. To suppose that historical prediction is possible even in principle marks a failure to grasp the essential nature of human affairs. The same mistake is made, though less spectacularly, when the past is deemed to be explicable in terms of "objective forces."[14]

4. Translation. Grasping meaning is always sizing up. It is not possible to learn how to extract meaning from discourse by learning a lot of rules. Of course, language is a rule-governed activity, but the crucial rules are syntactic. What meaning is to be conveyed by what phoneme depends on the whole context in which the latter occurs; this is what defeats the computer approach. In ruling out "The vodka was agreeable" as an equivalent to "The spirit was willing," we are not guessing.

5. Creation. This is in a way the converse of sizing up, but it is also straightforwardly sizing up, as the artist must constantly criticize his own effort as he goes along. He must work out a context in which the parts have the significance desired, and this is in no wise the mere addition of parts to parts. Painting is not putting down

14. However, this is not to deny that determinism in some version may be true. See Chapter VI.

one daub, then another, then another; composing
poetry is not writing the first word, then the second,
and on to the end, though a Martian would detect
nothing else going on. Though Michelangelo said that
he merely chipped away the superfluous marble to
release the statue inside, and, in a way, that is what
sculpture is, still deciding which bits are superfluous is
sizing up and not a rule-governed activity at all.

6. Mathematical proofs. The mathematician has to
develop a strategy of proof; this is a sizing up process.
Where what is to be dealt with is a set of axioms specifying purely formal operations to be performed upon a
given set, a state of affairs by no means characteristic of
mathematics in general, the computer is in its element
and can grind out theorems and proofs ad lib. Yet even
where this ideal is attained, as in *Principia Mathematica*,
there remains a difference between the computer and
the human approach. Some of the theorems will be
interesting; some will not. The machine cannot make
this distinction. And assuming, *per impossibile in excelsis*,
that it could, a distinction would still remain, for the
machine would prove everything and throw away the
uninteresting results whereas the human mathematician goes straight—pretty straight—to the interesting
ones. Even the eighth grader doing his geometry home-
work does.

7. The hunter pursuing his quarry. This and its
complement, the quarry fleeing the hunter, are the
primitive sizing up situations. A few plants, the most
ingenious being the Venus's flytrap, have developed a
sort of digital computer approach to capture, though not
to pursuit. But clever as the flytrap's triggering device
is, it is fundamentally different from the frog's snapping
at a moving fly. The frog's reaction is not a reflex, and

the frog can modify his behavior, learn, as the flytrap cannot. Every mobile animal, no matter how lowly, can learn[15]—even the one-celled; no plant can. And learning always involves sizing up. This is true even where the learning is the acquisition of a so-called conditioned reflex, for there must be some point in the process at which the two stimuli are jointly perceived and the animal sizes up the situation as food-in-the-offing, or whatever.

Hunting is something that machines may be fairly good at, as developments in military hardware show. However, machine hunting, like machine tennis playing or flying, differs in gross characteristics from the correlative animal activity. The heat-sensitive missile is pointed in the general direction of the aircraft it is to destroy, and off it goes, kept on its target by its servo devices, undeterred by countermissiles (unless hit by one) or anything else. At times, military tacticians have aimed at reducing the foot soldier to a similar automaton; but more enlightened generals encourage advancing in short rushes, taking cover, assessing risks, changing targets as opportunities arise, and similar distinctively animal responses.

8. Realizing that what you have done is harmful/admirable, and feeling remorse/pride. Though it is trivially true that a nonsentient being could not feel remorse or pride, neither could it size up and grade its own performances in the requisite way.

9. Construction of argument. Here we have a curious circumstance. If we knew of digital computers only that they were machines capable of drawing all the logical consequences of input data, we might naturally

15. Perhaps this is not true of some parasites, but it must have been at some time in their evolution.

suppose that they would set out arguments for us. But no. Lawyers, politicians, judges, historians, scholars, even mathematicians and logicians may find computers useful in many ways as ancillary devices; but there is no question of their actually producing the arguments with which these people deal. No one has even suggested that they might. If a computerophile would let a computer write a chapter of his book, that would be a coup indeed. However, perceiving the way premises go together in an argument to lead to a conclusion is like sizing up a field of battle or arranging the sequence of thoughts for a poem: a holistic activity not reducible to the independent bits on which a computer must feed and out of which it must make its constructions.

10. Getting confused. The concept of confusion has no application to digital computers. Strictly speaking, a computer cannot err, though it can have wrong input data or components not functioning according to its design specifications. Sentient beings, on the other hand, are liable to both error and confusion: error from wrong data, as in the computer; confusion, when the sizing up process does not go through. One can fail at it, as at anything else one tries to do. The ability to err or fall into confusion may not be a source of pride, but it is a genuine ability for all that.

11. Distractability is another defect from which machines are exempt. It is related, however, to an important advantage of sentience, namely, monitoring systems that prevent the sentient being from coming to grief through focusing attention so narrowly as not to notice peripheral menaces.

12. Emotion, perhaps another defect of our virtues. Some moralists may admire the computers of the labs, which laugh not, neither do they weep, nor run ineffectually in circles. But the emotionless machine in its way

reminds us of the truth that reason by itself can never be the source of any action or forbearance. Though the great-souled man recognizes that few things are important, he knows as well that the end of emotion would be the end of sentience and of life. We are emotional beings for good biological reasons, which there is no need to be ashamed of. Talk against the emotions betrays a wrong conception of reason, as being essentially opposed to emotional expression instead of being simply different.

13. Moralizing. This is neither a matter of discovering or pointing out a set of facts nor simply of feeling but of appraising or sizing up, which is the tertium quid alongside of fact finding and feeling. But morality is an elusive conception, and this is not the place for the long hunt.

14. Religion.

15. Science. The crucial step in scientific inference, the formulation of a hypothesis, is paradigmatically a sizing up.

16. Philosophizing. It would never occur to a machine that man is a machine.

Here are some more verbs that indicate sizing up:

account	beware	diagnose	justify
accuse	blame	divine	manage
acquit	camouflage	envy	mean
adapt	choose	evaluate	meditate
administer	comprehend	explain	mind
advise	conclude	extenuate	perceive
alarm	consider	fear	plan
allude	contrast	grasp	read
appraise	convince	guess	recognize
appreciate	debate	hide	see
apprehend	define	identify	understand
assess	deliberate	inspect	value
audit	design	judge	

One may want to ask whether there is any human or even animal action that is *not* a kind of sizing up. Well, this activity is so pervasive (there can be scarcely a moment, for instance, when we are not judging something, in a broad sense of "judge") that it may be doubted whether there is anything we do into which sizing up does not enter in one way or another. Nevertheless, there are activities in which it is not central, for example, shot-putting, weight lifting, and other exercises of sheer strength; repetitive actions such as simple dances; drills of all sorts; much musical instrument playing; washing clothes (but not dishes); swimming; games of chance; and typing. In general, anything that we can do without attending to it, that we can do "in our sleep," as it were, does not essentially involve sizing up. This may seem to imply that something that requires a sizing up initially can, with repetition, become something that does not require sizing up, as, for example, driving a car through heavy traffic, playing standard chess openings, or recognizing Socrates coming down the street. However, it is probably better to say of such cases that the sizing up has become routine so that it is carried out quickly and effortlessly, for sizing up can be very rapid, as shown by the capability of animals to respond in a flash to complicated emergencies. The actions in question cannot be done in a thoroughly automatic way but require some minimal attention, as brushing the teeth ordinarily does not. A Gestalt is still presented, and its significance still has to be appraised even though on the umpteenth occasion the appropriate appraisal does not have to be pondered.

These examples will have suggested that sizing up is

antithetical to following rules. This is correct, but we must be careful of the sense of "rule" involved. It is logically impossible for one who does not know the rules of chess, or who is determined to flout them, to be sizing up a situation in chess. Where rules define an activity—"constitutive rules," as Searle calls them[16]— not only must the agent behave in accordance with them in order to be engaging in the activity, he must do so with knowledge of the rules and with the intention so to behave.[17] Constitutive rules, however, do not ordinarily prescribe what in particular is to be done.[18] They leave open, or indeed create, a range of possibilities from which the agent may or must make a selection. Other rules, the prescriptive ones, do not constitute the conditions under which an activity must be carried on but lay down definite procedures to be followed if certain ends are to be pursued: Add one tablespoon of baking powder for each cup of flour; prune dead limbs all the way back to the trunk; form the plural by adding s. Directions are rules of more particular application: Solder the capacitor C147 to the terminal T14; turn left for Wilkes-Barre. A command or order is a direction addressed to a specific person for execution at a specified time with a penalty for noncompliance. Following

16. See J. R. Searle, *Speech Acts* (Cambridge University Press, 1970), Chapter II.

17. A spy in an exotic land watches people playing some game, of which he knows neither the point nor the rules. However, he notes that they make certain kinds of moves and not others. This serves him in good stead when the Tyrant forces him to participate. By making the moves he has noticed and avoiding those he has not, he may manage to pretend to play the game. But he does not appreciate the game situation, and he makes neither strategic nor tactical decisions, though in the context of deceiving the Tyrant, he is sizing things up in a very subtle manner.

18. There are exceptions, for example, forced moves in chess or rules for dancing the minuet.

rules and directions, obeying orders and commands, are not matters of sizing up.[19] However, rules can always be broken, directions disregarded, commands disobeyed. This is a logical point; something like a rule, but literally unbreakable, would be called a law or natural regularity whereas there is nothing that is just like a direction or order except in being of necessity followed or complied with. It is not a rule that people die, nor can they be given directions for dying or orders to do so.[20] Hence, to talk of machines as following rules, directions, or orders is an anthropomorphic metaphor. It is convenient to talk of giving a machine the direction or instruction to multiply seven by four and extract the square root, but we cannot literally do this any more than we can order a brick to fall from our hand. A machine performing operations in a sequence is not following rules but operating in accordance with laws.

All the same, we are justified in speaking of someone who meticulously and unimaginatively carries out directives as behaving mechanically, for undeviating regularity is the hallmark of the machine, which cannot size things up and modify its course accordingly. Mere following of rules is behavior where, as is said, you don't have to think.

Nevertheless, the learning of skills consists mostly in finding out how to proceed according to rules. Thinking takes time and effort and does not always come out successfully. Therefore, thought, like dentistry, aims at

19. Unless the rule or direction is a rule to size up, as it may be: to look both ways before crossing the street or to reconnoiter the enemy position.

20. Directions can of course be given for killing oneself or for dying in a certain manner, say heroically, but not for just plain dying. "Die, thou base villain!" is not an order but an optative or a statement of murderous intent.

abolishing the need for itself, as far as possible. The discovery of a pattern, procedure, or algorithm is the result of sizing up, an intellectual achievement that obviates the need for any more such triumphs in that particular field. This fact generates the paradox of the dullness of Utopia: If the dreams of the great intellectuals from Plato to Skinner were turned into reality, the resultant world would have little place in it for intellectuals. For once the Pattern of the Way Things Go has been grasped and the Pattern of the Way Things Ought to Be has been imposed, there is nothing left for the intellect to do. Once the randomness of the environment is conquered, there is no need or scope any longer for the exercise of the sizing up faculty that it called into being.

I have presented these examples of sizing up and the discussion of some of its aspects before attempting any formal consideration of the notion, in the hope that it has enabled the reader to grasp for himself this by no means mysterious or novel notion. Now it is time to attempt an explicit account, if not a formal definition.

The most concise way to express what I mean by "sizing up" is that I intend it to be a translation of the Greek verb *noein* and the cognate noun *nous*, especially as they are used in Homer and other archaic literature.[21] "Insight" comes closer than any other English word to this notion. However, as it lacks a cognate verb and is used in a quasitechnical way by philosophers and psychologists with whose views I would not wish my own

21. In the *Iliad* the verb *noein* almost always means notice, descry, or (especially) interpret the significance of. Heraclitus exactly captures my sense in his remark that "Polymathy does not teach a man to have *noos*" (Fragment 40, Diels) —in computer terms, vast storage capacity does not add up to pattern recognition. See the studies of *noein* by Kurt von Fritz in *Classical Philology*, 1943, 1945, and 1946.

to be confused, I prefer to use the term "sizing up," which may be defined more or less formally in the following way. Sizing up a situation involves:

1. Picking out, within the situation, certain features as relatively distinct;
2. Of these distinct features, recognizing some as more important than others and than those not distinguished (the background);
3. Of these important features, apperceiving the static or dynamic whole which they compose;
4. Relating this apperceived whole to one's interests;
5. Finally, sometimes, raising the question of what one is going to do about the situation.

Comments:

To 1. The world is presented to us through our sense receptors as a differentiated field. However, the differentiations that are given consist only of shades of color, motions, and (perhaps) differences of distance in the visual field; differences of pitch, timbre, and loudness in the auditory field; what we may vaguely call textures in touch; and various tastes and odors. Sorting these out into individuals and other features is our work. Whether the unit of our consideration is to be forest, grove, tree, branch, or leaf is not dictated by nature but decided by us according to our interest of the moment. When we are listening to talk in a language we know, we can group the phonemes into words and phrases, though an oscillograph record might not show gaps at those junctures.

To 2. Animals can seldom afford to gaze, listen, or sniff idly. When they inspect the world, they do so in keeping with preexisting interests. Some are standing orders, so to speak. The rabbit must constantly be on the lookout for the ferret, and the ferret for the rabbit. Some are special and temporary, as when we go gathering mushrooms. A perceptual field at any given moment will comprise a background of elements not differentiated by the perceiver, a foreground of elements both differentiated and selected for further scrutiny (zoomed in upon, as it were), and intermediate elements, for this is a continuum, not a dichotomy. But the relative status of the elements may change at any time, as the scrutiny proceeds. Those elements comprising the foreground must be described as "important." Herein is the beginning of value.

To 3. This might be called the sizing up proper. I say "apperceive" rather than "perceive," for it is of the essence of sizing up that the whole is not only perceived but also "united and assimilated to a mass of ideas already possessed, and so comprehended and interpreted."[22]

To 4. The relating of the whole to interests is not to be thought of as something done only after steps 1 to 3 have been finished; indeed, it would be a mistake to suppose that this listing represents a temporal sequence at all. In particular, our interests largely determine what will be foreground and what background. Among the interests to which the whole may be related, that of curiosity is important in some animals such as cats and men.

To 5. The first four elements of sizing up are integral to every instance, including every apprehension of meaning; for example, you accomplished all of them in reading the previous sentence, as you can easily verify. It would be somewhat artificial, however, to say that in

22. Concise Oxford Dictionary.

every such case one considered what to do. This element of sizing up, therefore, cannot be said always to occur. Nevertheless, mention must be made of it because sizing up is primarily a process directed toward practice. One may not consider what to do about every sentence in a book, but when we take account of larger units, the relation to practice becomes apparent even here.

Sizing up is a rational activity. Indeed, it is the rational activity par excellence, notwithstanding that we share the capability of it with dumb animals, insects,[23] and amoebas. Reasoning, whatever else it may be, is basically the preparation for an appropriate conscious response to a situation. It means taking account of all relevant information and putting it together so as to apprehend the situation as it really is. This point deserves emphasis. Sizing up can be expressed in a proposition to which the terms "true" or "false" can properly be applied. It may be just as true that the situation is threatening as that the cat is on the mat. The verification, however, may be more complicated, consisting not in noting the presence of some distinct objects in a definite relationship, but in effectively re-sizing up. A man who says that a situation is threatening can give reasons for his judgment; he may point out the approaching tiger, for example. That there is a tiger fifty feet away, that he has noticed me, that he is hungry, that neither I nor anyone else around has a weapon effective against tigers—from these premises one cannot *deduce* the proposition "I am threatened." All the same, those statements add up to a set of good reasons for

23. Bees seem to be good at it, ants less so.

accepting it.[24] Or recall John Wisdom's famous example of the woman remarking to another woman trying on a hat, "My dear, it's the Taj Mahal."[25] Perhaps one would prefer to call such a comment apt rather than true, but, on the other hand, the notion of truth may just stretch that far. Likewise, the metaphysician sizing up the universe and declaring it to be atoms moving in the void is right; his colleague, who avers that it is a system of floating ideas, is wrong. Sizing up, when done rightly, reveals what is the case. It can be the case that a hat is the Taj Mahal.

We have already noticed that the constructing of arguments is a sizing up activity. Formal logic, sometimes helpful for deciding whether a putative reason really is so and for exploring the implications of statements, cannot of itself generate reasons. Nor can "induction." Sizing up is the process whereby we reach general conclusions from particular premises,[26] though it is not the method whereby they are proved. Sizing up does not prove anything.[27] All the same, it is by sizing

24. One could, of course, construct a deductive argument by adding a premise to the effect that "Any situation containing these features: tiger, hungry, in vicinity, having noticed one, absence of effective weapons, is threatening." But this does no good because it only puts us back where we were, for our reasons for accepting this proposition are precisely our sizing it up that way. Programming a computer to flee when encountering this concatenation of circumstances would not protect it against assault by lions.

25. From "The 'Logic of "God," ' " in John Hick (ed.), *The Existence of God* (New York: Macmillan, 1964).

26. See my paper "Against Induction and Empiricism," *Proceedings of The Aristotelian Society*, 1962. Some instances of what is here called "sizing up" are there referred to as "getting a hunch."

27. But philosophers are apt to forget that comparatively few things need proof. The vast majority of our beliefs, though not necessarily the most important of them, not only do not need proof but do not even admit of that notion. It makes no sense to suggest that we prove that a certain hat is the Taj Mahal.

up that we get both our premises and our strategy of
proof.

In sizing up, values originate. All data are created
equal, which is why an automaton cannot size up; it has
no way to pick out the important items and find out
what they add up to unless it has been instructed to do
so in this one particular situation. Nor can the automaton do any more with a fact than register it as a fact.
Animals, on the other hand, pick out what is important,
what is interesting, that is, what impinges on their
interests—usually vital ones. Without this capability,
nothing would be better or worse than anything else,
for this classification presupposes a prior estimation of
importance and interest.

Most philosophers recognize that values have to have
this kind of origin.[28] The only alternative is a conception
of value as something already built into the fabric of the
universe and merely passively recognized by animals,
as we recognize mass and velocity. But that is a logical
muddle. If things were good and bad in this sense, then
after we had discovered what the values belonging to
the things were, we would be in the predicament of
having to start all over again to find out which ones we
ought to pursue and which to shun. We cannot simply
ignore our vital interests.[29]

To point to this fact is not to imply that values are

28. Including those who make good and evil dependent on the will
of God, hence on His sizing up.

29. "*In* [the world] there is no value, and if there were, it would be
of no value." Wittgenstein, *Tractatus* 6.41. See also M. Schlick,
Problems of Ethics (David Rynin, tr.; New York: Prentice-Hall, Inc.,
1939) Chap. 5, Sec. 7.

"subjective." The assessment of a situation in terms of the interests that I bring to it may be made rightly or wrongly. Someone wiser than I may properly criticize my evaluation even if he does not share my interests. A certain degree of "relativism" may be consequent, but who would deny that a man's being eaten by a tiger is bad for the man but good for the tiger? The interests with reference to which the question "What ought I to do?" is to be asked need not be exclusive or selfish. They may be any interests I happen to have, and if I do not have some that you think I should have, there is nothing to bar you from reproaching, educating, propagandizing, or browbeating me into sharing them. Interests reside in animals—there is nowhere else for them to live—but that does not mean that the individual animal can be interested only in the fate of its own skin and the contents thereof.

No natural language, as far as I know, has built into its verb structure a distinction between the so-called moral and nonmoral "oughts." We can see why. What you ought to do is what best satisfies the interests in terms of which the sizing up has been done. Those interests may be of various types—for individual or group survival, satisfaction of curiosity, self- or group aggrandizement, glorification of God, promotion of universal brotherhood. A man is said to be acting morally accordingly as he is responding to the demands of certain social interests, and immorally if he is not doing so when he might be and when others prefer that he should. Even to say that much is to mislead, as it suggests that there can be some neat definition of the sphere of morality, which is a very mixed bag. As this is not the place to work out a theory either of morality in particular or of value in general, I leave the topic.[30]

How do we size things up? The reader who wants an answer to this nuts and bolts question must go to the brain physiologists and learn about their exciting work, of which I cannot give here even a summary. Fascinating and important though it is, it has no direct bearing on the question with which we are concerned. This may seem paradoxical, for the question is "What is sentience?" and I have subscribed to the answer that it is identical with brain processes. So it must seem that the question becomes, "What are those brain processes that you say are identical with sentience?" But for our purposes it does not matter what the detailed answer to this question may be, as long as it is of the form "Matter in motion" or "Energy exchanges"; it might, for example, be "Flow of the animal spirits through little tubes," and the philosophical position would be just the same. To put the point in another way, to the question "What is sentience?" ideally the answer would be "It is brain processes X, Y, and Z." The brain physiologist one day may tell us what X, Y, and Z really are. But whatever they are, this kind of answer presupposes that sentience is brain processes. This statement is not self-evident

30. But some applications to aesthetics may be suggested. Sizing up, like all exercises of vital functions, is intrinsically pleasurable. However, it is difficult to savor it, so to speak, when it is directed to practical ends. Much of art provides nonpractical situations for sizing up, like play, but in more concentrated form because the spectator, unlike the player, is not called upon to do anything overtly. And yet the very highest of human faculties are exercised.

From this standpoint music has a just claim to the title of highest of the arts, for it demands that the listener perceive how its intrinsically meaningless data comprise wholes with an entirely self-contained meaning. Music has meaning, but it does not mean anything. That is why we like it.

apparently since many people deny that it is true or even that it makes sense. I have tried to show in Chapter II that it does make sense. The next step, which is still occupying us, is the descriptive task of characterizing sentience.[31] Only when we know *what* sentience is will we be in a position to appreciate *that* (moreover) it is X, Y, and Z.—But we are all sentient creatures; we all know what sentience is already.—Exactly. Insofar as I tell you anything you don't already know, I am not writing philosophy. One can, however, know all the facts yet not see how they fit together.

Let us now return to the question "What is the use of sentience?"

As with so many questions in philosophy, this one does not sound hard; on the contrary, it has a simple-minded ring. It seems hardly different from, "What is the good of being alive rather than dead?" For death is just permanent deprivation of sentience.

I mention this, however, only to dismiss it because it is not the sense in which I am raising the question of the use of sentience although, as underlining the dependence of all values on sentience, it is not entirely irrelevant to the question more literally understood, which is here intended: What is the *use* of sentience, what can sentient beings *do* because they are sentient? Now, I think that if this question were put to the vulgar, the response would be along these lines: "Why, of course, we can act *consciously*, which is to say, we can do things after making up our minds to; we can weigh evidence,

31. Some people may suggest that this is the exclusive bailiwick of the psychologist. That is not so. Psychology is or ought to be concerned with theories to *explain* the facts of sentience which are known to all.

make our decisions and choices in the light of it—or against it if we're so inclined—and know what we're doing." This is at any rate a correct answer; it is equivalent to saying that sentient beings can size up and act accordingly.

Yes, but how does consciousness make this kind of activity possible? We must remember that sentience does not *do* anything; it cannot be an agent.

Again, the simple reply is that in order to size up we must be able to look things over, and sentience enables us to do this. In the field of consciousness, the things are displayed so that we can scrutinize them and apperceive their relations. It is an obvious fact that sentient creatures can size things up. If we were assigned the task of devising a sizing up machine, *one* solution, in principle, would be to make it sentient.

Here there is need to be very careful, lest we fall into the trap of thinking of sentience as providing us with an inner picture that is then examined by *ka*, soul, mind, or ghost in the machine. Data from the world, in the forms of photons, air vibrations, and the like, impinging on the sense organs, set up disturbances in the nervous system that travel into the innards of the brain. We think of them as constituting, as it were, signals from the outer world; and it is tempting to suppose that these signals then have to have an interpreter, a "brain-writing reader."[32] This is, of course, ridiculously wrong.

But exactly what is ridiculous about it? We laugh because it is so patent an instance of the propensity to solve a problem by duplicating it. If there were a little

32. Daniel Dennett, *Content and Consciousness* (London: Routledge & Kegan Paul, 1969). p. 86ff.

man reading the brain writing, we could raise the same questions about what went on in his little brain, and so ad infinitum. And yet we cannot tolerate a picture which terminates in unread signals.

The solution is to give up the notion that brain events are signals making up some kind of internal representation of the external world. The brain processes are not a display to be scanned—they *are* the *scanning*. The display is there all right—it is just the world. *We* are the little men, looking at the display, reading not brain writing but the writing "out there." So the "model" of a man reading off from a display and taking appropriate steps turns out not to be ridiculous—not even a model, but the literal truth.

To put the point in another way, let us again suppose that we are given the problem of designing an all-purpose machine, one that is going to be put out into the environment with certain needs and with only the instruction to keep itself going. How would we go about this design problem? At least some necessary conditions for the working of the machine are clear. It would have to be able to gather information about the environment, which means that the machine would have to receive signals from the environment and interpret them relatively to its needs, such as "the fuel station is over that way." And these signals will have to be marshaled in some orderly fashion, just as you can't take a photograph by merely exposing a photographic plate to the light; you have to put it into a camera, the function of which is to order the incoming light. So, likewise, our all-purpose machine will have to have apparatus for ordering the signals. Then there has to be further apparatus for interpreting the signals in meaningful

ways; this is the sizing up faculty. Finally, there has to
be executive machinery for initiating appropriate
responses.

But all this is just a way of describing *us* looking at the
world and deciding what to do about it. There is no
internal brain display on the other side of the sense
receptors. There is no need for any; these organs simply
marshal the incoming signals in an orderly way so that
we (that is, our brains) can take account of them. What
we, the "little men," look at, and hear and smell and
taste and touch is no ideal duplicate of the real world
but that very world. Sensations are brain processes.
Sensations are not inspected; they are inspectings.

Let us take our bearings. The argument so far has
gone this way:

1. *Either sensations are necessarily brain processes, or they
   are contingently brain processes, or sensations and brain
   processes are distinct.*

This is an exclusive and exhaustive alternation of the
possibilities.

2. *But sensations are not contingently brain processes.*

For the notion of "contingent identity" is not coherent.
This was argued in Chapter II.

3. *Nor are sensations distinct from brain processes.*

By well-known arguments, here assumed to be cogent.

4. *Therefore, sensations are necessarily brain processes.*

This is to be read as a statement of class inclusion. Only
some brain processes, at best tentatively identifiable in
the present state of brain physiology, are sensations.

5. *Now, sentient beings can size up.*

This is obvious from experience.

6. *Any material device organized exactly like a sentient being would be sentient and could size up.*

The Frankenstein Axiom.

7. *Discrete state machines cannot size up.*

The argument of Chapter IV.

8. *Therefore, any device that was not sentient but was capable of sizing up would be neither a discrete state machine nor organized exactly like a sentient being.*

From 6 and 7.

I wish I could show the converse of 5, viz., whatever can size up must be sentient. 8 is weaker than that, allowing the logical possibility of some device that could size up, well enough, let us say, to win sometimes at Turing's imitation game, without being actually sentient. I cannot imagine how there could be such a device, but this psychological disability does not constitute an argument. It is solacing, however, to reflect that no one else can imagine it either; that is to say, if discrete state machines are incapable of sizing up, no one has the slightest notion how to proceed on the way to the creation of artificial intelligence. For nothing could be said to be intelligent that was not capable of sizing up.

VI

FREEDOM

The news, if it is news, that human beings are neither computers nor imitable by them, will come as a relief to some of us. Why? Mostly on account of our feeling that if machines can do what we do, then we are only machines ourselves, or so close as to make no difference. And the depressing thing about being a machine is that it does what it does because it has to. Machines are not free.

However, to rejoice at deliverance from machinehood would be premature at the present stage of the argument. All we have shown so far is that an artifact which came close to duplicating human (or, in general, animal) behavior could not be discrete state and would be either conscious or built on principles entirely unknown. Leaving aside this latter monster—let us call it the Mystery Machine—of which we can say nothing, not even whether or not it is logically possible, we are still far from having dug a chasm between the human and the artificial. On the contrary, the Frankenstein Axiom stands as a bridge set up in advance over any future excavation. Nevertheless, I hope to justify an inference from the nondigitalizability of human behavior to the reality of freedom.

The first step is to show that there is something to show, for many philosophers have maintained that the so-called problem of freedom (free will, free choice) is merely a muddle, neither demanding nor admitting any speculative solution but requiring only closer attention to the meaning of words. According to this line of thought, freedom signifies absence of compulsion, and since it is wrong to think of the laws of nature as compelling anything, we act freely all the time except when overwhelmed by alien force—as indeed machines do too. In the classic formulation of Hobbes, water runs downhill both freely and of necessity, at one and the same time. There are no external impediments to its motion, so it runs freely; it can't do otherwise, so it runs of necessity.

There may be something in this doctrine and its subtle descendants; but it is not sufficient to quiet the qualms that give rise to the problem. Water makes no choices, so what it does is not felt to be relevant to our kind of freedom. We are not free in the relevant sense unless we can make choices that can go either way. But when we do things that the world at large necessitates our doing, then we have no real choice. If we are machines, all our actions are of this sort.

To make the nature of this uneasiness clearer, let us contrast two biographies that might be written about the same man. The first is of the usual type. It is more than a mere chronicle of events; it explains why the subject did the things he did. Beginning with an account of his ancestry, the characters and achievements of his parents, grandparents, and other near relatives; describing the home in which he was brought up, his education, notable influences, travels, reading, ailments, escapades; it goes on to his early career, recounting his opinions and attitudes and the reasons for them; the process of his selection as a candidate for Congress, his campaign, why he failed miserably; the motives that led him to make a second try; and so on. Now suppose the subject of this biography reads it and agrees that the explanations of his actions are correct and complete, that is, they show how the alternatives to the courses he took came to be ruled out. This agreement will not give rise to any feeling that his freedom had been in any way or degree compromised. This will be so even if the biographer is a brilliant writer who provides insights into motives of which the subject himself was hitherto unaware.

Compare another biography which could in principle be written if human behavior is digitalizable. It begins with a sort of wiring diagram of the infant, deduced from the chromosome organizations of the parental sperm and ovum. Then it states characteristic input-output functions (the machine table), including interior modifications according to the nature of the inputs and the elapsed time since conception, that is, as we would say, the effects of growth and education on character. Thereafter, the biography consists of a series of specifications of input conditions, and deduction of the output via the machine table. For example (much simplified), Input: visual scanning of stack of $100 bills; impulses from input are equally distributed to GREED and CONSCIENCE centers, activating 37 relays in former, 19 in latter; output from GREED center overrides that from CONSCIENCE, feeding into MUSCULAR INNERVATION SUBSECTION TONGUE AND LARYNX pattern for production of noises "All right, I'll give your bill a DO PASS."

It can hardly be doubted that the subject of a biography of this type, if he were convinced of its accuracy, would feel that he had been mistaken all his life in believing that he was *doing* things. Really he had only reacted to stimuli in a manner determined by his hered-ity and that summation of subsequent inputs which we call environment. In any case of so-called decision, the outcome was already deducible from the inputs and the machine table so that the process of deciding, as experi-enced, was a superfluity. There would be a difference of complexity, but nothing more, between a knee jerk and a decision to pull out of the Common Market. To put it another way, the data from which the outcome of the decision can be computed are there already before the decision is made. But then the decision making process is really automatic and in a way bypassable. In particular, sentience is dispensable, for the computation need not take it into account.

The game of solitaire affords a simple model of the situation. Where the rules admit no options, a playable card when turned up must be played and in one way only; it is easy to see that the outcome is settled[1] once the pack has been shuffled and cut. Moreover, there must exist an algorithm whereby one can determine the game's outcome by inspecting the order of the cards before they are dealt. So solitaire is a trivial game, owing what excitement it may have to lack of knowledge about what has happened already, as in watching election returns. The player *does* nothing. This is very different from bridge and poker, where the outcome depends to a high degree on the wits of the players in sizing up situa-tions on the basis of partial information and making their plays accordingly.

However, if human beings were digital machines, the difference between solitaire and poker would be only of complication. A computer that is to predict the outcome

1. Barring mistakes or cheating. But these may be ignored in the analogy since Nature neither makes mistakes nor cheats.

of a game of solitaire needs for its input simply the order of the cards. For poker, it will require the card order plus the condition of the players. But that condition too is just an arrangement, though enormously more complex. As the outcome of solitaire preexists in the cards, the outcome of poker preexists in cards and players. Both are in principle calculable in advance.

If we were like that, it would be hard to see how freedom could be anything but a delusion. We suppose that before we choose, we think things over, notice this and that, regard one thing as of more importance than another, inject our prejudices, and consider our short- and long-range goals, and we suppose that these operations make a difference. But if the computer can print out our choice by following a sequence of operations on data fed in before we begin our deliberations, and without itself deliberating, then our thinking things over does not make a difference; or, at most, it is the ineffective epiphenomenon of the unconscious sequential moves between input and output. Computers are not free; ergo, if we are computers, or like them, we are not free either; indeed, there is no such thing as freedom.

For the freedom that we are concerned with is neither more nor less than this: that conscious deliberation should make a difference in the way things come out. *An action is free if* [2] *its full explanation requires reference to the agent's conscious deliberation.*

2. I do not say "and only if" because routine or automatic actions may be done freely without being deliberated about. The statement can be converted into a definition of freedom by adding some clause such as "or could have been modified or prevented by the agent in such a way that the modification or prevention would have required for full explanation reference to the agent's conscious deliberation." But this is cumbersome and unnecessary for our present purpose, which is to investigate the question whether there is at least *a* class of free actions.

Thus, the problem of freedom is not merely what used to be called a pseudo problem, a confusion to be cleared up by purifying the language. If belief in freedom involves a belief that the explanation of what we do cannot be complete if it omits all reference to our deliberations—and it is hard to see how anyone could deny that this belief is so involved—then the reality of freedom must be problematical for at least some philosophers, including those who hold that the digital computer is an adequate model of human behavior.

Now we know that the digital computer cannot model our behavior. And if we are partisans of freedom, that should be a relief to us, for we can eliminate one reason for denying freedom which would be conclusive if it were true. But of course the conclusion that we *are* free does not follow immediately from the fact that we are *not* computers. We must go back to the beginning.

"A full explanation of John Doe's performing the action A on the occasion O can be given without reference to his conscious deliberations." From our charac-

---

Comments on this characterization of freedom:

(1) In being positive, it accords with our intuition that to act freely is to do more, not less, than to act *simpliciter* (which is not to say that the more is another *action*).

(2) We must not simply say that an action is free if it is done as a result of deliberation, for that would be to define freedom into existence. Obviously, we deliberate, and obviously, actions result (in some sense of the word "result") from the deliberations.

(3) To say that the full explanation of an action requires reference to the agent's conscious deliberation is not to imply that the deliberation caused (produced, effected, brought about) the action. What it does imply will be discussed in the text.

(4) Taken as a definition (with the added clause), this formula preserves the gist of the familiar contention that one acted freely if one could have done otherwise. For what the proponents of that view intend is not something absurd such as a putative violation of causality, but merely the action's dependence on deliberation, that is, if the deliberation had taken a different course, the action would have too.

terization of freedom it follows that if every sentence of this form—let us call them Laplace sentences[3]—is true,[4] then we have, *pro tanto*, no reason to think that freedom is other than illusory; but if even one is false,[5] then freedom exists. If none is known to be false, but the falsity of such a sentence cannot be ruled out on logical grounds alone, then freedom is at least a logical possibility.

Assuming that we know what conscious deliberations are or, at any rate, what counts as a description of the process, hence how to distinguish accounts referring to conscious deliberations from those that do not, the only obscure term in the Laplace sentence is "full explanation." The notion intended is this: A full explanation of an occurrence is a true account that answers the question, "Why did that occurrence happen (rather than not) as it did (rather than in some other way)?" The full explanation, then, will be a set of true statements, each

3. "We ought then to regard the present state of the universe as the effect of its antecedent state and the cause of the state that is to follow. An intelligence, who for a given instant should be acquainted with all the forces by which Nature is animated and with the several positions of the entities composing it, *if further his intellect were vast enough to submit those data to analysis*, would include in one and the same formula the movements of the largest bodies in the universe and those of the lightest atom. Nothing would be uncertain for him; the future as well as the past would be present to his eyes. The human mind in the perfection it has been able to give to astronomy affords a feeble outline of such an intelligence. . . . All its efforts in the search for truth tend to approximate without limit the intelligence we have just imagined." Laplace (quoted in Arthur Eddington, *New Pathways in Science* [Cambridge University Press, 1935], Chapter IV). Emphasis supplied.

4. Assuming, of course, that John Doe actually performed the action A on the occasion O and consciously deliberated about it.

5. Disregarding trivializations, such as making the "action" A consist in a conscious deliberation.

one of which is relevant but not identical to[6] the explicandum and which jointly entail the explicandum.[7]
However, another condition must be added, that the
explanation be full *enough*. For example, in accordance
with what has been said so far, the explanation of why a
house burnt down might be simply that gasoline-
soaked rags in proximity to the walls and curtains
were heated to a temperature of 800°, which temperature is sufficient to ignite them and the adjacent
materials, . . .; which is all right, but we want to know
how and why the gasoline-soaked rags were heated and
indeed how it happened that they were there at all. If
the answer is that John Doe the notorious pyromaniac
put them there, we may require to know the cause of his
aberration. Or we may not; it depends on whether we
are insurance investigators or judges or psychiatrists. A
philosopher might not be satisfied with anything less
than a tracing back to the protoplasmal primordial
atomic globule. The cutoff point depends on the pur-

6. Non-identity excludes trivialities. The relevance condition implies the impossibility of a formal criterion for full explanation, for
relevance is a sizing up notion. It is, therefore, impossible to produce a
list of conditions for relevance, for if this could be done, relevance
could be decided by a digital computer, which, consequently, would
be capable of sizing up.

7. The contention that the relation of explicans to explicandum
must be logically watertight and not merely "empirical," requires a
defense which would be beyond the scope of this work. For a
summary statement see my paper "Against Induction and Empiricism," *Proceedings of the Aristotelian Society*, 1962. However, the issue is
not of great importance in this context, for the argument to be given
will work just as well whatever the relation between explicans and
explicandum is conceived to be, as long as it is agreed that full explanations are unique, that is, that there cannot be two or more valid full
explanations of the same occurrence.

pose of the inquirer. For our purpose, which is that of determining whether human beings ever act freely, an explanation of an action is full if it includes all relevant facts about the agent that have a direct bearing on the occurrence under investigation. Thus, we may need to know that the agent is a pyromaniac but not the etiology of the condition; that he is blue-eyed, but not the chromosomal characteristics of his grandparents.

Questions may also be raised about the depth and detail required if an explanation is to count as full. Would the explanation of Wellington's victory at Waterloo be full if it correctly assessed the contribution of every battalion? or company? or individual soldier? or every movement of every soldier? The notion of "full explanation" is exposed to the same perils as that of the complete map.[8] However, such problems need not delay us, as our concern is only with necessary conditions for full explanation, not with its actual production.

However, while the complete map must represent every feature of the topography, the full explanation can omit irrelevancies and otiose consequences. The explanation of why the cake failed to rise need not mention aberrancies in the oven temperature if the fact is that the cook forgot to put baking powder into the batter, for if the baking powder is omitted, the cake will not rise no matter what the oven temperature may be. Nor in explaining why the house burnt down is there need to say anything about the intensity of the light produced by the fire.

To return to the Laplace sentences: Now in our ordinary explanations we do frequently make reference to

8. Which, tarpaulin-like, would cover the entire territory it mapped. See Renford Bambrough, *Principia Metaphysica* (Cambridge University Press, in press), Chapter I.

conscious deliberation. How else could we explain a decision, or anything depending on it, than by giving the reasons? So the prima facie case against Laplace sentences is strong, being the same as our ordinary intuition that on many occasions we act freely.

The defender of Laplace sentences cannot deny that we do explain behavior as we do, by referring to reasons and choices. He has to say that they are irrelevant or otiose, the explanations in which they figure being in consequence bogus or at least misleading. No matter what we may think, all we can *do* is move our muscles. But muscle motions are physical events; and all physical events are unambiguously determined by their physical causes, which in turn were determined by their causes, and so on backward ad infinitum or to the beginning of things, if there was one. Therefore, the full explanation of my action can consist of an account of the relevant distribution of energy before the action, which, together with the laws of motion, provides the data for calculating the positions to be successively occupied by my muscles, that is, what I do. Just as from the ordering of cards in the pack and the rules of solitaire, the state of play at any time is determined, so from the original shuffling of the atoms and the laws of motion all their subsequent positions are determined. The full explanation of behavior, therefore, can bypass the conscious deliberation. The form is simply "Input to John Doe being such and such, and his internal state being so and so, his output was thus and so." What fills up the blanks is an account of energy distribution. No terms from the vocabulary of deliberation need be used. This is not to say that the account of deliberation could not be given; the Laplacian will admit or rather insist that from the physiological story the exact description of the process of deliberation can be inferred. Inferred, however, as a consequence of events in the nervous system, an accompaniment of them, as light is of the heat of combustion, and just as little executive. That is what epiphenomenalism means, that the conscious trappings of action do and perhaps must emerge from the physical events, which, however, would go on just the same if there were no such side issues.

That the possibility of this kind of full explanation eliminates freedom will become clearer if we think of it as shifted in time so that it is a prediction. We can always do so since the logical forms of physical explanation and prediction are the same. What the Laplacian says is that if he had enough antecedent information and a fast enough computer, he could grind out the prediction of what I am to do before I do it. But if it is possible in principle for an outsider to know what I am going to do before I do it, then my doings are already in the cards, I am swept along in the river of fate, or whatever metaphor you like; at any rate, my deliberation is not effective; hence, I am not free.

This simple argument gets its force from seeming to be a direct consequence of the principle that every event has a cause, that things happen in regular, that is, explainable and predictable ways. Perhaps that principle has exceptions, and philosophers from Epicurus to Heisenberg have essayed to save freedom by denying that human behavior is subject to strict mechanistic causal regularity. However, it is easy to see that this desperate move achieves nothing or even makes matters worse. For whatever freedom is, it certainly is not randomness. If my action has no cause, then it is not explainable at all, hence not explainable by reference to my deliberation, hence not free. Not constrained either, just crazy. And this must be the consequence on any plausible definition of freedom. The reason why the Swerve of indeterminacy has seemed even for a moment to offer a way out is perhaps that its proponents hope it will afford enough looseness, play between the parts of the world machine so to speak, to make it possible for something else—the Self in its role of rational executor—to alter the way things are going to go, no matter how they have gone up to the crucial point. If we have got to the situation where my finger is on the trigger and the muscles are tensing, we may suppose the causes that necessitate my pulling the trigger are already in operation so that the effect is inevitable. But if there is a Swerve, there never is total inevitability, so there is just the possibility that my Self, or whatever we are to call it, whatever constitutes my rational side, gets another chance, so to speak, to push the atoms in the other direction. But, of course, this suspenseful story is nothing to the purpose, for if the Self is a material atomic structure, the whole problem is merely moved back onto it: are there Swerves within the Self, and if so, then what? Whereas if the Self is not a material structure, then the whole system has been abandoned.

Thus, neither determinism nor its denial seems to offer any scope for freedom. Nor will it even help to abandon materialism, for although we can postulate an immaterial soul and endow it with whatever properties we please, our concern is with human actions, that is to say, bodily motions, which are wholly within the material realm, however spooky their causes. And yet our intuition that we are free is as strong as any of our fundamental convictions, for example, our belief in the existence of a world independent of our thinking. So it is no wonder that philosophers have sought to show that free will is a pseudo problem, hence need not be met head on.

But let us look once more at the crucial premise of the argument against freedom. That is the statement that all Laplace sentences are true—that every human action can in principle be explained without reference to conscious deliberation. Now inasmuch as no human action has ever in fact been explained in this way, and perhaps none ever will be, we may well wonder why anyone should venture so astounding an assertion, much less take it as obviously true. The Laplacian's reply is that it is certain because it is merely a special application of the general principle of causality, that motions of matter are explainable or predictable in terms of the laws of motion and initial conditions of the bodies involved. You collect the data about how big your objects are, where they are, and how fast they are moving; plug them into the equations of motion; turn the crank; and out come the coordinates for the new positions. The motions of the planets are in fact calculated that way;[9] so in principle it can be done for my motions too. The fact that I am "self-moved," so to speak, complicates the issue but does not alter it fundamentally. For the data needed are all of the same type: mass, velocity, charge, spin. The laws are expressed in ordinary differential equations. At no point is there need or even opportunity for the insertion of any information concerning what I am thinking, even though, no doubt, the calculations could be interpreted to express my

9. It is not so simple a matter as is sometimes supposed. The calculations of the moon's motions alone are staggeringly complicated and, before the advent of high-speed computers, could barely be kept ahead of the actual motions being predicted. Even with computers the most extensive calculations that have been done for the major planets cover only the period 1653–2060, a mere instant in astronomical time.

thoughts; for example, from such and such motions in brain and stomach, it could be inferred that I feel hungry and am plotting to steal a loaf of bread.

This then is the Laplacian view of nature and man, impressive in its austere simplicity and still perhaps thought to be that to which anyone who regards himself as a materialist is committed.

However, let us ask impudently how the predictions envisioned are in principle to be made. To this, Laplace's explicit answer is that they are within the province of analysis, which amounts to saying that the calculations can be made in a finite number of operations performed on a finite store of information, in short, that they can be made by the methods of the digital computer. But if the conclusions of the preceding chapter are correct, we know that it is, in principle, impossible to predict or explain all human behavior in this way. If the digital computer could predict how I am going to size up a situation, it could itself size up that situation; the prediction would be such a sizing up, and that we have seen to be impossible. To continue a previously used example, if the machine is to predict whether I will laugh upon hearing a certain string of words for the first time, it must know whether I am going to regard them as funny. But wit and humor lie in the significance of words, not in any of their physical characteristics taken either singly or together. Therefore, the machine would have to be able to grasp significance, which a discrete state machine can by no means be made to do.

Does this then prove that we are free? No, but (I submit) it comes close; and from here on, the road should be downhill.

To prove that we are free, according to the characterization of a free action as one whose full explanation requires reference to conscious deliberation, it will suffice to show that (1) there is at least one action that is fully explicable, and that (2) that action cannot be fully explained without bringing in conscious deliberation. Now it is not difficult to satisfy requirement (1). Let the action be *my making the statement* "An action is free if it cannot be fully explained without reference to the agent's conscious deliberation." The explanation of that action is that in the course of writing this chapter, I needed to pin down the conception of freedom; I considered several alternatives; I concluded that one of them was better than the others, for reasons a, b, c, . . ., n; then I decided to make the statement; nothing prevented me from making it; so I made it. This is an explanation of the action. Moreover, it is a full explanation, for the last clause is entailed by the others.[10] "Because I decided to, and nothing prevented me" is always a full explanation of an action. If it seems disappointingly meager, that is because we are usually not interested in the explanation of actions per se but of decisions.[11]

The satisfaction of requirement (2) is now not difficult either even though it amounts to the proof of a universal negative.

Since I did in fact deliberate before I acted, and my deliberation is part of the explanation of my action, the only hope of giving a full explanation that dispenses with deliberation lies in bypassing it—showing that it is not fundamental. This means that if the alternative enterprise is to be successful, there must exist the possi-

10. Or, rather, by the two preceding clauses, "I decided to X and nothing prevented me" entails "I Xed." If anyone wishes to make it a bit more elaborate, for example, "I decided at time t_0 that I would X at or before t_1, and I did not alter my decision nor forget it in the interval t_0 to t_1, and nothing prevented me from Xing before t_1," so be it.
11. Usually but not always, for example in criminal law.

bility of predicting my action from antecedent data that do not include any reference to deliberation. And how might this be done? We have just seen that the Laplacian suggestion, which amounts to a proposal to feed the data into a digital computer, will not work. This leaves just two possibilities:[12] do it with a conscious device, or do it with a Mystery Machine.

A conscious device could conceivably predict my behavior by dealing with the data that will be input to me in the manner that I would deal with them, that is, by duplicating what we may call my characteristic sizing up parameters. Such a process would be another sizing up, a simulation of mine: The explanation or prediction would be by *Einfuehlung*. This, of course, would necessarily involve reference to my deliberation. It would be of the form, "When data D are input to Matson, his sizing up will be as follows: . . . and in consequence his action will be A."

If a Mystery Machine is possible, it will also produce a printout of my sizing up. The only difference from the conscious device will be the irrelevant one that we can hardly speak of *Einfuehlung* in connection with an unconscious artifact. But there will be the same reference to conscious deliberation on my part. Someone might demur from this proposition on the ground that the Mystery Machine need handle only physical data, that is, the physical events that are the inputs to my sense organs, plus the physical processes that occur in my nervous system when I size up; that the Mystery Machine proceeds in its mysterious way to process these data nondigitally until it produces its printout,

12. More accurately, one possibility and one possible possibility; for we do not know whether a Mystery Machine is even possible. But as we shall see, it does not matter.

without either itself being conscious (by hypothesis) or referring to my consciousness (for it refers only to brain events). The answer to this objection is that the brain events referred to *are* my conscious deliberation, by the identity theory (see Chapter II).[13] The identity theory, the doctrine of sizing up, and the reality of freedom are not isolated but form a unity.

Thus, freedom is real. The tasks remain, first, of showing that this statement is not Pickwickian but refers to what the plain man supposes he has; second, of examining a few implications of the fact.

I take the ordinary criterion of freedom to be just this: I am free to X to the extent that my deliberate decision to X makes a difference as to whether I do X; moreover, I choose to X or not according to the reasons I have and (when there are reasons both ways) according to their relative importances as I apperceive them. Although we speak of compelling reasons, we do not at all mean thereby that when all the reasons are in favor of Xing and none against, we are forced to X as by the lash of the slave driver. On the contrary, I am never so free as when, in Luther's sense, I "can't do otherwise."

The argument of this chapter entirely vindicates this freedom. Luther acted freely—there could be no explanation of his behavior that failed to refer to the way

13. The philosophical reader will have noticed that the identity theory, at any rate in the strong version urged in this book, presupposes a thesis of extensionality, that contrary to the philosophy of Brentano, it is always in principle possible to replace an intentional sentence (for example, "I am afraid of ghosts") by a nonintentional sentence ("My brain includes such and such a dynamic structure") *salva veritate*. In my remarks about pain (Chapter V, p. 130) I have tried to support the possibility of such reduction. A much fuller speculative account is given by Daniel Dennett in *Content and Consciousness*. Fortunately, a formal proof of the requisite thesis has been produced in the very nick of time; see George Bealer, *A Theory of Complex Qualities* (Dordrecht: D. Reidel, 1976).

Luther sized things up. He defied papal authority because that was to him the action called for in the situation, and unlike John Hus *et al.*, he was not overwhelmed by physical constraints. If the brain physiologists of the future show that he stood firm because the
F-fibers in his pineal gland were unusually stiff and
unreactive, that will not be a rival explanation supplant-
ing the citation of reasons and rendering them super-
erogatory; it will be only a further explanation of what
Luther's reasoning *was*.

In this connection it is important to note that when we
say John Doe acted as he did because of reasons R and
S, we do not mean that R and S were the causes of his
action, the things that produced it or brought it about,
as worry and alcohol produced his ulcer. Reasons are
not things or even processes, much less brain structures. To cite reasons is to (try to) reproduce salient features of the sizing up that issued in the action. *I, you,
Luther* act; to specify the reasons for which we act is in no
way to climb further up a causal chain. Sometimes,
indeed, we are subject to causes; that is to say, our
behavior is explainable just in terms of what happens to
us; but on those occasions we are not agents but passive
transmitters of alien forces.

Keeping these points in mind, it is not difficult to
answer the question that goes back to Democritus and
Socrates, whether human behavior ought to be
explained teleologically or mechanistically—in terms of
purposes and goals or of pushes and pulls. The answer
is that the full explanation, which would be the same as
the full description, the citation of every relevant factor
in its structural context, would be both, and both at one
and the same time. It would, of course, be "mechanistic"; that is, it would detail the patterned interactions of

the material particles comprising the body because, to put it bluntly, there is nothing else there; whatsoever is done by whatever means is alteration in the pattern of energy distribution, and nothing else. But some of these processes, occurring within the skins of animals, are processes of sentience, including sizing up. Here and only here is the being of intentions and values. They are not shadows but real denizens of the world, inasmuch as they are identical with patterns of energy distribution. But they are known to us in a direct and unschooled way. We know what it is to have the intention of sailing to Hawaii and how it differs from the intention to sail to Tahiti, though we do not know, and perhaps never will, exactly what brain configurations they are and how they differ when regarded as such. Nor would we somehow have a deeper knowledge if we did. We want explanations in order to understand; understanding is one kind of sizing up; it is entirely appropriate, therefore, that the categories of understanding should be the familiar and natural categories of sizing up, where they can be.

When we explain in terms of goals, desires, purposes, wants, motives, and the like, we are making behavior intelligible by fitting it into a pattern. The purposive explanation is itself a sizing up. It is not a causal explanation; goals and the like do not push people, nor pull them either, in any nonmetaphorical sense. Conscious awareness of a desire to sail to Tahiti is, to be sure, an event in the brain, consequently causally linked to other physiological occurrences. Just how, we do not know, but we do know that the linkage cannot be anything so simple and straightforward as "the desire causes me to go and buy a yachting cap." The brain events that constitute sentience are (we can be sure) a tiny fraction of the whole number of brain events, and their causal link-ages are altogether hidden from introspection. We have no notion at all, in awareness, of how an itch is causally linked to scratching. The events that enter into aware-ness are so to speak only the foamy crests of the brain waves. Explanation of behavior in terms of them exclu-sively, then, would be inherently incomplete: While they can be formed into a satisfying pattern, they cannot of themselves exhibit the necessity of that pattern's being the only possible one. This fact is most of the reason behind the view of many philosophers that only purposive explanations allow for freedom—because there is enough play in them to preclude necessitation—and that insofar as causal explanation is admitted into human affairs, to that extent freedom is a lost cause. But it should be clear by now that this inference is but another confusion generated from the initial mistaking of freedom for randomness.

But here we cannot treat further the many and vast topics that open up when we begin to ask about the nature of explanation. The remarks just made should suffice for stating and dealing with a final objection that is bound to be urged against an account of freedom like the one presented in this book:

"You admit or even insist that it is, in principle, possi-ble to give a complete mechanistic explanation of human behavior. That means that whatever anybody does is in principle fully explicable by antecedent condi-tions and the laws of physics. So whenever anybody finds himself in the position of making a choice, a really super intelligence could figure out, from the state of the world before the agent made his choice, or even before he was born for that matter, what the outcome would be. We may concede you the point that the super intelligence would not make this prediction by programming a digital computer, but by *Einfuehlung*. What of it? The so-called agent is still necessitated to do what he does; and that is not freedom."

But what this amounts to is the reiteration that if someone knew all the reasons that influence me and what weights I give them, he would be able to duplicate my deciding; and if he could do it faster than I can, he could come up with the answer before I do. This is not necessitation in any but the innocent sense that the premises of a true explanation logically necessitate the conclusion.

Nevertheless, while the objection considered as a general attack on the vindication of freedom is of little force, it contains an important and valid point, the Spinozistic insight that individual men are but modes of nature and cannot be conceived independently of it. When I act, it is I who act and not somebody or something else. But what am I? I did not make myself, either in matter or in form. I am a hunk of organized matter, with many abilities, including the ability to develop new abilities. Highly characteristic of me—so essential that we might as well call it my Self—is the way I size up situations and act in them. That Self is not something static; it changes all the time, or, at least, it may, both from external impingements and from what I myself (the Self) decide to do with it. But the core, so to speak, is a given, from heredity and early environment. I cannot do or become just anything; there are limits, present in that core: obvious for eye color, unknown but nonetheless real for amenability to rational persuasion in the paths of virtue. To put the same point in different words: Freedom means that our behavior is not (always) mere passive response to stimuli; we are (sometimes) active and even truly creative in our dealings with the situations we find or put ourselves in. But the "we" who are active are ultimately products of nature, determined to be what we are by powers that antedate us. We are free, hence justly to be praised when we act rightly, justly to be censured when we behave atrociously, both because virtue is praiseworthy and vice blameworthy and also because praise and blame in the offing are elements in sizing up situations that may have considerable importance in our decisions. But the praise to which we are entitled or the blame we incur cannot be of a different kind from what we bestow on a well-wrought urn or a poorly wrought typewriter. Despite the reality of freedom, in the end the fault does lie in the stars.

Index

Action, 166, 182; moral, 155. *See also* Freedom; Free action
Adrian, E. D., 4
Aesthetics, 156 n.
Agency, 122
Analytical engine, 103, 114, 120, 132; question answering ability of, 112 ff; game recognition ability of, 105; language learning ability of, 106; imitation game with, 112. *See also* Computer(s); Imitation game
Analytic/synthetic distinction, 20 ff.; and a priori truth, 24 f.; and W-θ identification, 40, 50
Animals, 125-128; and value, 154
Apperception, 150 f., 158, 178
A priori/a posteriori distinction, 20; and necessity, 24 ff.
Argument, 143
Aristotle, 5, 9, 22, 23, 24, 46 n., 80
Armstrong, D. M., 6, 28, 52 n.
Art, 134; and recognition, 135
Artificial intelligence, 135, 161. *See also* Computer(s)
Artifact(s): and discrete states, 162; as a sentient being, 83, 162; definition, 80; limitations of non-conscious, 86; necessary conditions for, 79; sufficient conditions for, 79 f. *See also* Machine(s)
Austin, J. L., 45, 52 n.
Autocerebroscope, 54, 72, 73 n. *See also* Cerebroscope
Automaton, 154
Awareness, 57

Babbage, Charles, 103
Bambrough, Renford, 170 n.
Bealer, George, 178 n.
Behavior, human, 164, 172, 181. *See also* Computer(s); Explanation; Machine(s)
Behaviorism, 7, 55; crude, 69 f.
Belief, 87, 88 n.
Bergson, Henri, 137 n.
Brain, 101; as a computer, 101; brain process (the term), 10 f.; events, 159; functions, 118; processes, 159. *See also* Computer(s); Mental activities
Brentano, Franz, 178 n.

Causality: general principle, 174
Cause, 122; and reasons, 179; as distinct from agency, 122. *See also* Explanation
Cerebroscope, 70; as obstacle to identity theory, 72; correct description of, 70; functioning of, 70; perfected, 73; uses of, 73. *See also* Autocerebroscope
Choice, 163, 181; and freedom, 163
Chomsky, Noam, 106
Commands (orders), 147 f. *See also* Rules
Compulsion, 163
Computation, 165
Computer(s), Chap. IV *passim*; and addition, 90 f.; and computing, 89 f.; and the brain, 101; analogue, 101, 116 n.; digital, 103, 116 n., 127, 175; scanning, 132; and creation,